Revenue and Economics 2018

AA005505

Transportation Research Record: Journal of the Transportation Research Board

Volume 2672, Issue 4

ISBN: 978-1-5108-7731-3

Printed from e-media with permission by:

Curran Associates, Inc.
57 Morehouse Lane
Red Hook, NY 12571

Some format issues inherent in the e-media version may also appear in this print version.

Published by Sage Publications USA

Copyright© (2018) by Transportation Research Board of the National Academies
All rights reserved.

ISBN (Print) 978-1-5108-7731-3
ISBN 2018 Printed Set (All Issues) 978-1-5108-7735-1

Printed by Curran Associates, Inc. (2018)

For permission requests, please contact sagepub.com/journals-permissions

Additional copies of this publication are available from:

Curran Associates, Inc.
57 Morehouse Lane
Red Hook, NY 12571 USA
Phone: 845-758-0400
Fax: 845-758-2633
Email: curran@proceedings.com
Web: www.proceedings.com

Volume 2672 Number 4 December 2018

TRR
JOURNAL OF THE TRANSPORTATION RESEARCH BOARD

Contents

Revenue and Economics 2018

Articles

Article

TRR
JOURNAL OF THE TRANSPORTATION RESEARCH BOARD

Transportation Research Record
2018, Vol. 2672(4) 1–12
© National Academy of Sciences:
Transportation Research Board 2018
Article reuse guidelines:
sagepub.com/journals-permissions
DOI: 10.1177/0361198118784405
journals.sagepub.com/home/trr
SAGE

Development of a Transparent Framework for Pre-Procurement Evaluation of Public–Private Partnership Project Delivery Options

Patrick DeCorla-Souza[1]

Abstract

This paper reviews value for money (VfM) analyses conducted in the United States to gain an understanding of methods used in addressing key issues in VfM analyses. The paper shows that VfM analysis approaches are inconsistent. In many cases it is difficult to decipher the source of differences between delivery options. Discount rates are sometimes used in a way that reduces credibility of the results. The paper presents some ideas for how these key issues can be approached so that VfM analysis results are more credible, consistent, and transparent.

Public–private partnerships (P3) are long-term contracts between public and private sector partners that involve financing, design, construction, operation, and maintenance of a public facility by the private partner. Value for money (VfM) analysis is often used by public project sponsors in the United States to evaluate the appropriateness of P3 project delivery. Due to the complex financial modeling involved, VfM analysis is generally performed for public project sponsors by financial consultants.

The purpose of this paper is to review VfM analyses conducted in the United States to gain an understanding of methods used by State Departments of Transportation (DOTs) and their financial advisors in addressing key issues in VfM analyses. Based on this review, the paper offers suggestions for a credible, transparent, and consistent framework for analysis. The paper focuses on pre-procurement analyses because the intent of recent congressional mandates in the Fixing America's Surface Transportation Act (FAST Act) and the Moving Ahead for Progress in the 21st Century (MAP-21) Act is to help public project sponsors make cost-effective project delivery choices before the project is tendered for bid. It is at the pre-procurement stage that the delivery option choice is made. Although the paper also reviews one analysis undertaken after P3 bids were received, it is notable that the results of such analyses are often simply used to confirm whether the expected value was achieved by the selected delivery method. Post-bid VfM analyses seldom, if ever, would result in a decision not to move forward with the chosen delivery method. Furthermore, as actual P3 proposals are available at the post-bid stage, there is no need to estimate a "shadow bid," that is, an estimate of a bid that may be received in response to a request for proposals. Developing a shadow bid is usually the most difficult part of a pre-procurement analysis. There are fewer issues in post-bid VfM estimation, and those issues are a subset of the issues encountered in pre-procurement evaluation.

Pre-procurement VfM analyses are carried out to help understand whether P3 delivery would provide value, based on a combined evaluation of cost, schedule, and quality of service. A shadow P3 bid must first be estimated. This estimate is usually based on a variety of assumptions, which can be a source of debate. This paper seeks to assess whether there are ways to strengthen the assumptions and make the results more credible, simpler to explain to the public, and more useful for decision making about project delivery choices.

The paper first discusses the various types of P3 evaluation, focusing on VfM analysis methodology as defined in international guides (*1–3*) and FHWA's P3-VALUE tool (*4*). It then discusses key issues raised by VfM analysis methodology in the U.S. context. Findings with regard to these issues are then highlighted from a sample of VfM analyses performed in the United States. Finally, the paper offers some suggestions for use of a credible, consistent, and transparent framework to address the issues.

[1]U.S. Department of Transportation and Federal Highway Administration, Washington, D.C.

Corresponding Author:
Address correspondence to Patrick DeCorla-Souza:
Patrick.Decorla-souza@dot.gov

Overview of P3 Project Evaluation

Types of P3 Project Evaluation

There are three types of evaluation that may be used by public project sponsors in assessing project delivery options, namely:

a) Financial feasibility analysis, aka financial viability analysis;
b) VfM analysis; and
c) Benefit–cost analysis (BCA), aka economic efficiency analysis.

Financial Feasibility Analysis. Financial feasibility analysis focuses on how costs to the public agency compare with financial resources available to the public agency for delivery of the project either as a P3 or using conventional delivery. It may include consideration of project revenues such as tolls. Regardless of which delivery option provides more value, agencies still have to be able to afford the nominal amount of payments required throughout the life of the project. Debt capacity and other capital investment constraints need to be considered.

Value for Money Analysis. VfM is generally expressed as the estimated dollar or percent difference between the whole-life costs of a project *to the public agency* under conventional project delivery (called a public sector comparator or PSC) relative to its costs *to the public agency* under a P3. The projected risk-adjusted lifecycle costs and revenues under the two delivery options are discounted to estimate their costs in net present value (NPV) terms. The focus of VfM analysis is on costs to the procuring agency, in conjunction with qualitatively described factors such as earlier project delivery, budget certainty, and quality of service provided to facility users, or other non-monetized factors. Benefits and costs to facility users (e.g., travel time savings and costs of tolls paid by users) are not generally included in the quantitative analysis – thus higher toll rates that may be imposed under one alternative relative to another can appear to be "beneficial" to the public agency from the perspective taken in VfM analysis, whereas higher congestion delays to users when project completion is delayed are ignored in the quantitative assessment. In fact, a delivery method that delays project implementation may actually appear to be "beneficial" to the public agency financially, because the NPV of postponed construction costs is reduced by the discount rate.

Benefit–Cost Analysis. BCA is a comparable method to compare delivery options. The perspective taken with BCA is much broader than that taken with quantitative VfM analysis. *Societal* costs and benefits (i.e., not just those that accrue to the public sponsor) are quantified and monetized, including costs and benefits to users of the asset. BCA helps estimate the overall net benefits to society from making the infrastructure improvement. It can also be used to estimate costs and benefits resulting from earlier or delayed delivery of the improvement, for example, due to differences in financial capacity of the public agency. Finally, if reliable information on cost, schedule, and quality of service differences between the conventional and P3 options is available, BCA can be used to estimate benefits or disbenefits from P3 delivery itself, relative to conventional delivery.

Overview of Value for Money Analysis Methodology

In conducting VfM analysis, there are five cost components that are commonly estimated for each delivery option under consideration, as discussed below.

- Base costs: These are all lifecycle costs, including costs for rights-of-way, design, construction, routine maintenance, major maintenance, and operations. Risks transferred to the private sector (i.e., to concessionaires, or to contractors or subcontractors) are included in the price of their bids, whereas costs for risks retained by the public sponsor are commonly accounted for as contingencies as discussed under "retained risks" below. Uncompensated bid preparation costs are accounted for in the price of bids.
- Financing costs: Payments to financiers include financing fees, interest on debt (i.e., debt service cost minus amount borrowed), and return on equity (i.e., equity dividend distributions minus equity invested). The sum of base costs and financing costs for the P3 option is used to estimate the NPV of the P3 shadow bid price in an availability payment (AP) concession. In a toll concession, forecasted revenues are additionally considered in calculating the shadow bid price, which would be a concession fee if the NPV of revenues exceeds the NPV of costs, and would be a required public subsidy if they are lower than the NPV of costs. Note that the conventional use of the term shadow bid to mean the entire cost of the P3 option is inaccurate – there are additional cost components that must be included in the full estimate of the P3 option, as noted below.
- Retained risks: These are cost estimates for risks borne by the public agency that may occur in the design-build (DB) phase (accounted for as contingency set asides) or over the life of the project, including the operations phase, that is, revenue risk and lifecycle performance risk. Under DB delivery, most risks are transferred to the DB contractor; under a P3, most DB as well as long-term risks are transferred to the concessionaire.
- Ancillary public sector costs: These are public agency costs for procurement, including stipends provided to bidders (if any) as well as costs for monitoring and oversight over the project lifecycle, that is, for the

construction phase as well as the operations phase. Although ancillary cost differences between delivery options may be small relative to the magnitude of project costs, acknowledging these costs can make the analysis more credible to stakeholders.

- Competitive neutrality adjustment (CNA): CNA includes costs for insurance required to be obtained by a P3 concessionaire that would be avoided under conventional delivery due to self-insurance by the public agency. CNA may also include tax payments made by the P3 concessionaire or debt providers that the public sector would forego if the decision were made to use conventional delivery. Though foregone state taxes may be small, taxes forgone at the federal level can be significant, and may be of interest in national evaluations or in BCA.

The sum of the above five cost components for each delivery option, converted to NPV terms, provides the overall NPV of that option. In the case of a comparison between conventional delivery and a toll concession, the NPV of toll revenues to be received by the public agency from the project under the PSC are subtracted from the NPV of total estimated costs of the PSC. Regardless of the quantitative estimate of VfM produced by analytical tools, each option will always have non-financial or non-quantifiable impacts that will need to be considered in combination with the quantitative VfM estimate. For example, VfM may not directly value cost certainty or project acceleration under a P3. Therefore a qualitative assessment of non-quantified impacts is normally undertaken and this assessment is combined with the quantitative results in the decision-making process.

Review of the Literature on VfM Analysis Practice

Several studies have been undertaken over the past decade critiquing the state of the practice in conducting VfM analyses. These are reviewed briefly below.

In June 2006, the Canadian Centre for Policy Alternatives, British Columbia Office reviewed several case studies of VfM analysis produced by Partnerships British Columbia (Partnerships BC) (*5*). One of its key findings was that the studies were "subjective, susceptible to manipulation by vested interests, complicated and consistently withheld from appropriate public scrutiny." Further, the review found that "the discount rates applied to P3s greatly understate the current value of future P3 contract payments, making them appear less costly." The paper suggests that VfM analysis typically relies on the arbitrary quantification of risk transfer to make up for the higher cost of P3s in cash terms.

In 2009, Morallos et al. reviewed international VfM guides and surveyed U.S. transportation agencies on their use of assessments and tools to evaluate P3s (*6*). The international review found that the approaches to assignment of discount rates and valuation of risks varied. Partnerships Victoria (in Australia) uses a risk-free rate for "non-risky" cash flows and a risk-adjusted discount rate for "risky" cash flows. The UK uses a risk-free discount rate, but calculates the value of risks and incorporates them into the cash flows of each procurement option. Finally, South Africa's National Treasury and Ireland's Central Public–Private Partnership Unit use a single discount rate equal to the government's borrowing rate. Some agencies use VfM analysis as a tool in deciding which procurement option to use, whereas others use it as a bid selection tool after bids are received. In the United States, only five agencies had used some form of VfM analysis, including "market valuation." Texas used market valuation to evaluate potential revenue risk P3s. This technique is essentially a financial feasibility assessment that estimates the value of the P3 option by calculating the amount of upfront concession fee that a potential concessionaire might be willing to pay, or the amount of upfront subsidy that the public agency might need to provide to make a toll concession financially viable. Morallos et al. did not assess VfM studies conducted in the United States.

In 2011, FHWA (*7*) published a review of the state of the practice in VfM analysis, covering VfM methodologies and practices in the UK, Australia, and Canada, as well as a few states in the United States. The review concluded that British, Australian, and Canadian approaches are similar in measuring public sector costs, and all quantify risks and determine whether the risk reduction potential under P3 is worth the price premium charged. The review concluded that VfM assessment is both time- and resource-intensive, and quantifying risks is the most complex and difficult step but is crucial in determining the appropriateness of a P3.

In 2012, M. Rooney reviewed the VfM assessment conducted on the US 460 project in Virginia (*8*). The review pointed out that the base date for the comparison of NPV costs can make a significant difference, because Virginia Department of Transportation (VDOT) would incur fairly significant costs before financial close – which were not included in the US 460 analysis because the VfM analysis compared the delivery options at financial close. The review also noted that one key assumption was that a significant portion of the shadow bid revenue would be driven not by innovative use of technology or customer focus, but by the assumed ability of the concessionaire to raise toll rates more aggressively than the public agency could. In reality, charging higher toll rates is effectively an additional cost to the region and may not provide any benefits to the region.

In 2015, Barutha and Scheepbouwer compared VfM analysis approaches in Virginia and New Zealand (*9*). They note that New Zealand guidance states that it may be difficult or impossible to assign correct dollar values to all elements in the analysis, but is nonetheless useful in the evaluation of P3s (*10*).

Furthermore, in 2015, Gifford et al. examined the state of the practice in VfM analysis of highway projects in the

United States (*11*). A mix of pre-procurement and post-bid evaluations was reviewed, but the authors noted that "the paucity of VfM studies available online or from public agencies was surprising, and may suggest the sensitivity of many deliberations leading to the selection of a P3 approach." They found that the most commonly used discount rate was 5%, as a proxy for the state's long-term cost of capital. They note that in Australia, a risk-free discount rate is used for the PSC model, and a project-specific rate (accounting for project risks) is used for the P3 model.

Additionally in 2015, Boardman and Hellowell compared nine government units with documented VfM methodologies and identified four ways in which they favor the P3 option by making adjustments to cash flows and discount rates (*12*). They argue that, to be consistent with general public policy norms, the procurement decision should be made on the basis of social welfare (aka economic efficiency), not government cost minimization.

In 2016, FHWA released its P3-VALUE 2.0 analytical tool and guide and this was updated when version 2.1 was released in 2018 (*4*). Three advances are notable with this effort. First, the guide and tool allow for social welfare/economic efficiency evaluation in addition to conventional VfM financial evaluation. Second, the tool accounts for an additional component not commonly considered, that is, operations and maintenance (O&M) cost savings relative to the No Build case. These savings can differ among delivery alternatives if the schedule for project delivery differs among the alternatives. Finally, the tool and guide recognize that it is practically impossible to estimate the public sector's value of risks transferred to the concessionaire as the value ascribed to the transfer of these risks will vary based on the perspective of the individual decision maker. However, an estimate of the private sector's project risk premium can be made based on P3 financing conditions and does provide a market estimate of the price that would be charged by a concessionaire to take on those risks. At the tool user's option, the NPV of the private financing risk premium may be added to the PSC cost estimates as an approximate value of the long-term performance and revenue risks that would be retained by the public agency under conventional delivery.

Key Issues in Conducting VfM Analysis

Key issues that affect the results of VfM analyses relate to:

- Development of credible estimates of the five cost components, that is, base costs, financing costs, retained risks, ancillary costs, and CNAs;
- Selection of the base against which a P3 option will be compared;
- Selection of the discount rate to convert nominal cash flows to NPV terms, and uncertainty in the appropriate discount rate.

Each of the issues is discussed further below.

Base for Comparison

PSC is the term used to describe the base against which P3 delivery alternatives are compared. PSC estimates the present values of the cost and revenue cash flow streams of conventional delivery, including adjustments for risks, over the project's lifecycle.

PSC is generally the option that the public agency would use if it did not choose to use a P3. PSC could be design-bid-build (DBB), construction manager/general contractor (CM/GC), DB, design-build-finance (DBF), or even a long-term contract with public financing, such as design-build-maintain (DBM) or design-build-operate-maintain (DBOM).

The choice of the base against which a P3 option is compared will impact the magnitude of VfM estimated. The key questions are:

- Which method of delivery should be selected as the conventional delivery option? If DBB is the option selected as the base case, should other "intermediate" options such as DB or DBOM also be evaluated to assess whether another option other than a P3 would provide even more value than a P3 would?
- Could the PSC be reasonably expected to be delivered in the same time frame as the P3? Budgetary, debt capacity, or staffing constraints often make conventional delivery of a major project difficult or impossible for the public sponsor to achieve in the same time frame as a P3. Thus PSC delivery could be delayed relative to the P3 option. It may take much longer (sometimes decades) for project construction to be completed under a pay-as-you-go approach. If these timing aspects are input into the financial model, the model would discount the future costs heavily. Consequently, conventional delivery would appear to have a lower present cost simply due to the artifact of discounting. How should this issue be addressed?

Discount Rate

Present value is the base year value of a sum of money due in the future, obtained by applying a discount rate. The discount rate represents the "exchange rate" between present and future sums of money. It is the percentage by which future project costs and revenues are reduced for each year between the base year and the future year when the cash flow occurs. When considering delivery options for an AP concession, a high discount rate will tend to favor the delivery option that has more of its costs occurring further into the future. Thus, the discount rate can have a significant impact on the VfM estimate. But there is no consensus among analysts on the appropriate discount rate to use. The key questions are:

- What is the appropriate discount rate?
- How should uncertainty about the appropriate discount rate be addressed?

Cost Components

P3 costs could be vastly different from those under conventional delivery. Additionally, more aggressive tolling strategies under a P3 toll concession might bring in higher toll revenues, and access design innovations may bring in more toll payers.

The estimated magnitude of P3 differences relative to the PSC will have significant effects on VfM estimates. The magnitude of P3 differences is influenced by P3 "value drivers" such as:

- flexibility to innovate, integration of phases (scope bundling) and a whole-life perspective, which can reduce lifecycle costs;
- performance-based requirements, which can assure superior quality of service in the operations phase, and provide flexibility to innovate;
- private sector management, which can improve cost and schedule certainty and service quality; and
- transfer of both long-term performance risk and revenue risk (in toll concessions), which can provide budget certainty for the public agency, can attract and provide benefits to more users (albeit toll payers) in a toll concession, and can minimize costs if risks are allocated to the party in the best position to control and manage them at the lowest cost.

These value drivers are modified by agency and concessionaire skills, contract flexibility, political support for P3s, enabling statutes, competition, and project complexity and risk.

Given the variety of factors influencing the magnitude of P3 differences, it is difficult to estimate the cost, schedule, and service quality differences resulting from P3 delivery. P3 differences affect all five cost components of delivery options discussed above: base costs, financing costs, retained risks, ancillary costs, and competitive neutrality. The questions relating to each cost component are discussed below.

Base Costs

- How credible are the assumptions about cost differences under a P3?

Financing Costs

- Financing efficiencies, for example, the upfront public funding contribution available for each delivery option, will affect its financing costs, so will bond financing vs. sculpted debt. Since the financed portion of investment depends on the magnitude of the upfront public contribution, a higher upfront public contribution will reduce financing costs. To what extent do the assumed upfront public contributions differ between PSC and P3?
- Risk transfer, for example, the risk differential between a PSC funded with general obligation bonds vs. project revenue bonds (which have no recourse to the state treasury), will affect the financing cost of the PSC; and project risks transferred to a P3 developer will increase the premiums included in P3 financing costs, affecting the financing cost of the P3 option. Are the risk differences accounted for in the VfM comparison?

Retained Risks

- Do contingencies set aside by the public agency under PSC and P3 account for the value of all risks that would be retained – not just for the DB phase, but also for the entire project lifecycle? Cost estimate reviews undertaken for major highway projects jointly by FHWA and state DOTs account for risks in the DB phase but long-term risks are not assessed. The key question is: Are risk costs included in P3 financing risk premiums accounted for in the PSC, including lifecycle performance (i.e., O&M and major maintenance) risks, and revenue risks for a tolled facility?

Ancillary Public Sector Costs

- Are differences between delivery options in procurement costs and costs for oversight and monitoring accounted for in the analysis?

Competitive Neutrality Adjustment

- Are federal, state, and local taxes foregone under PSC (relative to the P3 option) accounted for?
- Assumptions regarding availability and use of a Transportation Infrastructure Finance and Innovation Act (TIFIA) loan and its terms can make a significant difference in financing costs, as these loans are subsidized by the federal government and have other advantages. Likewise, the federal tax subsidy for state and local tax-exempt debt significantly lowers financing costs for conventional delivery, and tax-exempt private activity bonds (PABs) significantly lower the costs for P3 delivery. Considering congressional interest in efficient expenditure of federal funds, should TIFIA subsidies and tax exemptions for government-issued debt and PABs be considered?

A Review of VfM Studies in the United States

We reviewed seven VfM studies to understand whether and how they addressed the issues discussed above. The VfM studies were obtained primarily from project websites. The studies were undertaken for three types of P3s: (1) toll concession (i.e., revenue risk P3s); (2), AP concession on a tolled facility; and (3) AP concession on a toll-free facility.

Although we found several post-bid VfM studies, few pre-procurement studies were available. We found no pre-procurement VfM studies for toll concessions. We therefore requested examples of such studies from a couple of states that have done toll concessions, but could obtain only one example of a pre-procurement VfM study report for a toll concession.

The studies were carried out in five different states for the following projects:

1. ***US-36*, *Denver*, *CO*** (*13*): US-36 connects the City of Boulder, Colorado to I-25, which provides onward connections to downtown Denver. The US-36 Express Lanes project reconstructed the four general-purpose lanes and added one high-occupancy toll lane in each direction along a 15-mi segment of US-36. The project also included accommodations for Bus Rapid (BRT) service and associated transit station improvements.
2. ***I-66*, *Northern Virginia*** (*14*): VDOT's Transform 66 project outside the Washington, DC Beltway will reconstruct and expand I-66 to provide:
   - two tolled express lanes in each direction, converting the existing high-occupancy vehicle lane per direction into an express lane and adding one new express lane per direction;
   - three general-purpose lanes in each direction;
   - high-frequency bus service with predictable travel times; and
   - direct access between the express lanes and new or expanded park and ride facilities.
3. ***I-70*, *Denver*, *CO*** (*15*): This project, called the Central I-70 project, proposes to reconstruct a 10-mi stretch of I-70 east of downtown, add one new express lane in each direction, remove the aging 53-year old viaduct, lower the highway on one segment, and place a 4-acre cover park over a portion of the lowered highway.
4. ***I-595*, *Broward County*, *FL*** (*16*): The I-595 Corridor Roadway Improvements project involved the reconstruction and widening of a 9.3-mi section of I-595 along with frontage road improvements and new interchanges, and the addition of three at-grade reversible express toll lanes (595Express) in the median of I-595. The reversible lanes operate in the direction of prevailing traffic, eastbound in the mornings and westbound in the afternoon and evenings.
5. ***Brent Spence Bridge*, *OH/KY*** (*17*): This project consists of a 7.8-mi segment of I-75 in the Greater Cincinnati/Northern Kentucky region. It carries I-75 and I-71 traffic over the Ohio River. The project seeks to address congestion, safety, and geometric problems on the bridge and its approaches.
6. ***Presidio Parkway*, *San Francisco, CA*** (*18*): The Presidio Parkway project replaced and expanded Doyle Drive, a 1.6-mi, four-lane east–west segment of Route 101 that serves as the access route to the Golden Gate Bridge in San Francisco. Doyle Drive did not meet current highway standards and was at risk of damage in the event of an earthquake. Caltrans adopted a two-phased approach to procure the Presidio Parkway. It packaged four of eight contracts into a single Phase 1 procurement using DBB. The Phase 1 procurement included a portion of the new roadway and a temporary bypass route. Caltrans selected a P3 option for the remaining four construction contracts (Phase II), as well as the long-term operation and maintenance of the full Presidio Parkway project for a period of 30 years.
7. ***Portsmouth Bypass*, *OH*** (*19*): Also known as the Southern Ohio Veterans Memorial Highway (State Route 823), this project will build a new four-lane, limited-access divided highway around the city of Portsmouth in Scioto County, Ohio. The 16-mi roadway includes 5 new interchanges, 19 bridges, and 16 ramps.

We summarize our observations on the above VfM studies in three tables. Table 1 covers the two toll concessions (US 36 and I-66); Table 2 covers the three AP concessions on toll facilities (I-70, I-595, and Brent Spence Bridge); and Table 3 covers the two AP concessions on toll-free facilities (Presidio Parkway and Portsmouth Bypass). Highlights are presented below.

Base for Comparison

Key findings were:

- DBB was the option selected as the base for comparison in only two studies (Presidio Parkway and Portsmouth Bypass). For the Presidio analysis, an "intermediate" option (DBOM) was also evaluated to assess whether another option would provide even more value than a P3 would, compared with the DBB.
- Two studies (I-66 and Portsmouth Bypass) recognized that the PSC might not be delivered in the same time frame as the P3, and could take several more years to complete due to financing constraints. Since this delayed PSC option would have congestion impacts causing delays to users, the economic costs of the user delays were accounted for in the I-66 study, but not in the Portsmouth study.

Discount Rate

Approaches taken in the studies varied:

Table 1. Summary of Findings from VfM Studies for Toll Concessions

Issue	US 36, Denver, CO (post-bid VfM)	I-66, Northern Virginia
Base for comparison	PSC is DB.	Two alternative PSC scenarios were estimated: PSC-1 was DB in the same time frame as P3; PSC-2 involved a 20-year construction period using pay-as-you-go financing. User costs with the delayed PSC-2 are accounted for in the PSC-2 cost estimate.
Selection of discount rate	5% rate applied to costs based on agency's cost of funds; 14% rate applied to revenue, based on equity rate of return.	5% rate used for both costs and revenues (but risks were accounted for separately through adjustments, as noted below).
Cost components:		
• Base and financing costs	PSC uses tax-exempt bond financing. Actual bid was used for P3 option, as the VfM evaluation was post-bid.	P3 differences were not quantified, but accounted for qualitatively.
• Retained risks	PSC was not adjusted for potential cost overruns and revenue risk, but a sensitivity analysis was conducted to evaluate the impact of long-term operations and maintenance/reconstruction and rehabilitation (O&M/R&R) risks and revenue risk. Revenues were discounted at a high rate (14%) to account for revenue risk. The 14% discount rate reflects the equity rate of return expected by the winning P3 bid.	Long-term O&M/R&R cost risks were quantified. Similarly, adjustments for revenue risk were quantified.
• Procurement and oversight costs	Not mentioned in report.	Not mentioned in report.
• Competitive neutrality adjustments	None made.	None made.

Note: VfM = Value for Money; PSC = Public Sector Comparator; DB = Design-Build; O&M = Operations and Maintenance; R&R = Reconstruction and Rehabilitation; P3 = Public-Private Partnership.

Table 2. Summary of Findings from VfM Studies for Availability Payment Concessions on Toll Facilities

Issue	I-70, Denver, CO	I-595, Broward County, FL	Brent Spence Bridge, OH/KY
Base for comparison	PSC was DB, and DBOM was considered in addition to P3.	PSC was DBF. Two scenarios for budget availability were considered for PSC and P3 construction payments.	PSC was DB.
Selection of discount rate	Not mentioned in report.	5% discount rate was used based on the state's long-term cost of capital.	5% discount rate was used based on the state's long-term cost of capital, except for non–risk adjusted revenue cash flows, which were discounted at 8.5% based on equity internal rate of return (IRR).
Cost components			
• Base and financing costs	Both DBOM and Design-Build-Finance-Operate-Maintain (DBFOM) were estimated to have lower costs compared to DB.	Two budget availability scenarios were considered. With higher budget availability, financing costs were reduced under both PSC and P3, but P3 costs were reduced more due to reduced reliance on higher cost P3 financing.	Costs for PSC assumed to be higher by 5% (optimistic) and 15% (conservative). Senior debt service coverage for PSC was 2.75 and 2.25; for P3 AP it was 1.25. For TIFIA, coverage was 1.48 for PSC and 1.25 for P3 (AP).
• Retained risks	Contingency amounts were estimated for each option. Long-term risks were considered in qualitative terms.	DBF pricing assumed 5% cost overrun and 10% contingency for construction; and 20% cost overrun and 10% contingency for O&M. For P3, 5% risk contingency was added to P3 construction pricing.	Risk costs during project development, construction.
• Procurement and oversight costs	Not mentioned.	Construction engineering and inspection cost estimated at 15% of const. cost for DBF and 5% for P3 (split 2% private/3% public).	Not mentioned.
• Competitive neutrality adjustments	Not considered.	Forgone state corporate taxes considered in PSC.	Not considered.

Note: VfM = Value for Money; PSC = Public Sector Comparator; DB = Design-Build; O&M = Operations and Maintenance; R&R = Reconstruction and Rehabilitation; P3 = Public-Private Partnership; DBOM = Design-Build-Operate-Maintain; DBFOM = Design-Build-Finance-Operate-Maintain; DBF = Design-Build-Finance; TIFIA = Transportation Infrastructure Finance and Innovation Act; IRR = Internal Rate of Return.

Table 3. Summary of Findings from VfM Studies for Availability Payment Concessions on Toll-Free Facilities

Issue	Presidio Parkway, CA	Portsmouth Bypass, OH
Base for comparison	The PSC was DBB, but DBOM was also considered.	PSC was DBB procured in three phases over 13 years.
Selection of discount rate	Used 8.5% discount rate, but also performed sensitivity analysis with lower rates. Break-even discount rate was 4.4%.	Used discount rates of 3.5% and 5%, resulting in a VfM difference of 1% and 13% respectively.
Cost components:		
• Base and financing cost	DBB has no financing costs, due to pay-as-you-go funding assumed for construction. DBB costs assumed higher costs for construction, and for the O&M phase due to effects of deferred maintenance. Financing costs for P3 were estimated as being negative, due to use of a discount rate higher than the P3's weighted average cost of capital.	P3 option incorporates 5% efficiency savings in capital and routine maintenance costs. Higher nominal costs were assumed for PSC due to exposure to high inflation rates over the 13-year construction period. PSC does not use TIFIA loan, but P3 does, reducing its cost of finance.
• Retained risks	Risk differences during the construction phase are accounted for with contingencies, but risks in the operations phase do not appear to be specifically addressed.	The quantitative analysis did not account for retained risks, including potential reductions in retained risk under P3.
• Procurement and oversight costs	Procurement transaction cost and oversight costs were estimated for both PSC and P3.	Not accounted for separately.
• Competitive neutrality adjustments	Forgone federal and state taxes were accounted for.	Not considered.

Note: VfM = Value for Money; DBB = Design-Bid-Build; PSC = Public Sector Comparator; O&M = Operations and Maintenance.

- Most studies used a discount rate of 5% approximating the state's borrowing rate.
- The Presidio study used a discount rate of 8.5%, which was higher than the weighted average cost of capital estimated for the P3 option. As a result, the NPV of P3 financing costs produced a "saving" compared with conventional delivery.
- Two studies (Presidio and Portsmouth) performed a sensitivity analysis to test the effect of alternative discount rates on VfM.

Cost Components

Key findings are discussed below by cost component.

Base Costs

- One study (US 36) was post-bid, so the actual bid cost was used.
- Another study (I-66) did not attempt to estimate impacts of P3 value drivers on costs, but instead simply recognized the possibility of these potential cost differences in the qualitative evaluation that accompanied the quantitative assessment.

Financing Costs

- The Presidio Parkway study assumed that the public agency's upfront contribution from its budget for the PSC would be more than three times the upfront contribution under the P3 option; as the PSC construction cost was funded from the agency budget entirely (pay-as-you-go), there were no estimates for financing costs.
- The I-595 study provided for the same upfront public agency contributions under both the PSC and the P3 options, and demonstrated that higher upfront public contributions would decrease overall PSC and P3 costs by reducing costs for financing. As financing is more expensive under a P3, a higher public contribution had the effect of increasing VfM. (A limitation under P3 is that the project needs to include sufficient private financing to provide an incentive to the concessionaire to continue to perform and not abandon the project if unforeseen costs arise, making the public agency worse off in the case of early termination.)
- For the Portsmouth study, a TIFIA loan was available only to P3, reducing its financing costs. Furthermore, debt service coverage ratios (DSCR) were much lower for P3. It is unclear what revenues were being pledged in the publicly financed alternative, causing its higher DSCR.

Retained Risks

- Most studies recognized risks in the DB phase, using estimates of contingency amounts that would be included in the pricing of the contractor's or the public agency's contingency – or both the contractor's and public agency's contingency – set aside.
- Some studies, but not all, specifically recognized and quantitatively estimated PSC lifecycle performance and revenue risks in the operations phase.

Table 4. Example of VfM Results Table in a VfM Study

Input or assumption	DB model (US$ million)	AP model (US$ million)
Public contribution/subsidy (construction)	(950)–(685)	n.a.
Public contribution/subsidy (operations)	n.a.	(814)
Net toll revenues less APs	n.a.	(804)
Excess cash flow after AP concession term	n.a.	146
Excess cash flow after debt service	556–619	n.a.
Retained risks	(194)	(108)
Rights-of-way	(61)	(61)
Tolling operations costs	(449)	(449)
Net project value/(costs)	(1,035)–(833)	(1,276)

Note: DB = Design-Build; AP = Availability Payment; VfM = Value for Money.
Source: Ohio Department of Transportation and Kentucky Transportation Cabinet (*17*).

Ancillary Public Sector Costs

- The Presidio study accounted for differences in costs for procurement as well as for monitoring and oversight over the project life cycle.
- One study (I-595) estimated construction engineering and inspection costs as 15% for the PSC and 5% for the P3 (split between public and private).
- Most of the other studies did not break out ancillary costs as a separate line item, so it is difficult to say whether they were included in the analysis and whether and how they differed between delivery options.

Competitive Neutrality Adjustment

- The Presidio study accounted for both state and federal taxes foregone under PSC.
- The I-66 study and the I-595 study accounted for state taxes only.
- The other studies made no adjustments.

Presentation of Analysis Results

We also reviewed how results of the VfM analyses were presented to assess whether the sources of differences between the delivery options were transparent and could easily be understood. A typical results table is provided in Table 4. We found it was often difficult to decipher the source of differences between delivery options by reviewing the analysis results. Often, the relationship of the results to input assumptions and impact on the various cost and revenue components could not be understood without reading the full report. Financing cost differences in particular were hidden.

Suggestions for a Framework That Addresses Key Issues

Based on the above findings, this paper presents a framework to address the key VfM analysis issues and present results in a way that clearly and transparently shows the differences between the PSC and P3 option for each cost and revenue component. Table 5 presents the framework using a hypothetical project. The PSC is compared with two P3 options: an AP concession and a toll concession. The framework reflects a societal perspective, but the same framework could be used to perform analyses from other perspectives by ignoring those line items that are irrelevant for the perspective desired. The framework is discussed below in the context of Table 5.

Base for Comparison

The framework considers whether the PSC might in fact not be delivered in the same time frame as the P3 – whether it could take several more years to complete due to the public agency's staffing, funding, or financing constraints. Since a delayed PSC option would cause congestion delays to users and other economic impacts, these economic costs are accounted for in the framework. Accounting for economic costs also neutralizes the effects of discounting on estimates of the present value of delayed expenditures.

A new line item (cost line 6 in Table 5) shows the potential value of additional social benefits from a delivery option that accelerates the start of operations or provides increased social benefits in some other way (e.g., by attracting more users or providing a better quality of service to users). For example, the P3-VALUE 2.1 tool allows high-level calculation of such benefits with minimal required traffic volume inputs, using default assumptions for many parameters such as roadway lane capacity, relationship of travel time to volume-to-capacity ratios, and value of travel time (*4*). Line item 6 in Table 5 reflects an assumption that the project's start of operations under conventional delivery will be delayed.

Selection of Discount Rates

Using a discount rate of 5% (which is common practice) along with the current approximately 3% interest rate for a TIFIA loan would produce negative financing costs, which would seem to be unreasonable to the public and could

Table 5. Value for Money Analysis for a Hypothetical Project from a Societal Perspective: Present Values (US$ million)

Component	PSC	P3-AP	P3-toll concession	Explanation
Revenues				
Revenue to agency	+US$1,000	+US$1,000	US$0	P50 estimate
Revenue to concessionaire	US$0	US$0	+US$1,000	P50 estimate
Costs				
1. Base costs (including risks transferred to subcontractors)[a]:				
Design-build phase	–US$1,000	–US$1,000	–US$1,000	Both PSC and P3 use DB procurement
Operations phase	–US$200	–US$200	–US$200	P3 costs include private procurement and oversight costs
Base costs subtotal	–US$1,200	–US$1,200	–US$1,200	
2. Financing costs[b]	–US$50	–US$300	–US$600	Fees plus debt service and equity distributions, less financed amount
P3 Shadow bid	n.a.	–US$1,500	–US$800	Base + Financing costs – toll revenues received by concessionaire
3. Retained risks[c]				
a. Design-build phase	–US$100	–US$100	–US$100	Both PSC and P3 use DB procurement
b. Operations phase	(?)	(?)	US$0	Risk transferred to P3 is included in its financing cost as a project risk premium
Retained risks subtotal	–US$100	–US$100	–US$100	
4. Ancillary costs[d]				
a. Design-build phase	–US$20	–US$20	–US$20	Lower oversight costs compensate for higher procurement costs under P3
b. Operations phase	–US$20	–US$5	–US$5	
Ancillary costs subtotal	–US$40	–US$25	–US$25	Majority of procurement and oversight costs are transferred under the P3
5. Competitive neutrality adjustment[e]:				
Insurance	US$0	+US$10	+US$10	P3 option credited with insurance costs to account for public agency self-insurance
Federal debt subsidies	–US$20	–US$20	–US$20	Accounts for TIFIA loan and muni bond interest subsidies
State taxes	US$0	+US$15	+US$22	P3 option credited for state taxes paid by concessionaire
Federal taxes	US$0	+US$100	+US$150	P3 option credited for federal taxes paid by concessionaire
Subtotal of all adjustments	–US$20	+US$105	+US$162	
6. Social welfare benefits[f]	US$0	+US$100	+US$100	Assumes accelerated start of operations under both P3 options
Net project value (after revenues)	–US$410	–US$420	–US$663	PSC excludes value of all long-term risks, and P3–AP excludes value for revenue risk
Differences relative to PSC		+US$10	+US$253	The decision on VfM will rest on whether higher P3 costs are worth the risk transfers not accounted for in PSC (lifecycle risk only for P3–AP, both lifecycle and revenue risk for P3 toll concession)

[a]Base costs: Line item 1 assumes no cost differences between conventional (DB) delivery and P3 delivery.
[b]Financing costs: Line item 2 transparently shows the financing cost components of the PSC and the P3, so that decision makers and the public can clearly see the magnitude of the financing cost differences.
[c]Retained risks: Line item 3a shows the public agency's contingency set aside for the DB phase. Line item 3b shows the value of long-term risks transferred to a P3 concessionaire, which would be retained by the public sector under PSC. For PSC, both lifecycle performance and revenue risk are retained. For the P3–AP concession, only revenue risk is retained. The question marks indicate that a value is not estimated, but may be considered qualitatively in combination with quantitatively assessed VfM differences shown at the bottom of the table.
[d]Ancillary costs: Line item 4 shows the estimates of the procurement and oversight costs by phase (DB vs. operations phase). They transparently show the assumption that the overall effect of the P3 option will be to lower these ancillary costs.
[e]Competitive neutrality adjustments: Line item 5 shows the different sub-components of the CNA for insurance, federal debt subsidies, and federal and state taxes. The P3 estimate reflects all four adjustments, whereas PSC reflects only federal debt subsidies for TIFIA and tax-exempt debt.
[f]Social welfare benefits: Line item 6 shows the potential value of additional social benefits from a delivery option that accelerates the start of operations or provides increased social benefits in some other way (e.g., by attracting more users or providing a better quality of service to users).

reduce credibility. To avoid such credibility issues, one approach might be to use a discount rate that does not incorporate project-related risk. One example is the federal long-term borrowing rate (i.e., the TIFIA interest rate). Use of a project risk-free borrowing rate can avoid distortions in the estimates of financing costs. It would also seem to be reasonable from a societal perspective.

Cost Components

Base Cost Differences. Cost efficiency is an important impact of P3 delivery. To the extent practicable, analysts should attempt to quantitatively estimate impacts of P3 value drivers on costs and schedule using ranges of potential P3 differences (low vs. high). If credible information on these ranges is not available, these differences should be considered qualitatively in a trade-off analysis that compares the VfM differences of the delivery options with their qualitatively described differences. Table 5 (cost line item 1) assumes credible information on P3 differences is not available, so cost differences between conventional (DB) delivery and P3 delivery are not estimated, and will need to be considered qualitatively.

Financing Cost Differences. The framework (see cost line item 2 in Table 4) transparently shows the financing cost components of the PSC and the P3, so that decision makers and the public can clearly see the magnitude of the financing cost differences.

Retained Risks. The framework includes separate lines (3a and 3b) for DB phase vs. long-term (operations phase) risks. Long-term risks that are transferred to the concessionaire include lifecycle performance risks and revenue risks (for toll concessions). These are retained in the PSC and need to be accounted for. Although it is practically impossible to credibly estimate the cost of long-term retained PSC risk (as the value ascribed to these risks will vary based on the perspective of the individual decision maker), the project financing risk premium based on P3 financing conditions does provide a market estimate of the price that would be charged by a concessionaire to take on those risks. These values can be used to help stakeholders and decision makers understand the price they would be paying to transfer those risks. The value of long-term risks transferred to a P3 concessionaire under a P3 but retained by the public sector under PSC are accounted for in line 3b. The question mark indicates that a value is not estimated for this hypothetical project, but may be considered qualitatively in combination with quantitatively assessed VfM differences shown at the bottom of the table.

Ancillary Costs. The framework accounts for public costs for monitoring and oversight over the project life cycle. For the PSC, in addition to DB phase procurement and oversight costs, these costs include costs in the operations phase for pavement inspection, procuring maintenance contracts, oversight of maintenance contractors, and so forth, which would be the responsibility of the concessionaire under a P3. On the other hand, procurement costs can be higher for a P3, and should therefore be accounted for. Table 5 (cost line item 4) transparently shows the assumption that the overall effect of the P3 option will be to lower these ancillary costs for the hypothetical project.

Competitive Neutrality Adjustments. When taxes or federal debt subsidies differ between P3 and PSC, these differences should be accounted for, from a societal perspective. Insurance differences should also be accounted for. The framework allows the analysis to account for taxes, insurance, and federal debt subsidies. The P3 estimate reflects all three adjustments, whereas PSC reflects only federal debt subsidies (for TIFIA and tax-exempt debt). Note that insurance and tax adjustments are shown as credits for the P3 options instead of costs for PSC.

VfM Trade-off Evaluation

After accounting for project revenues (shown at the top of Table 5), the NPV of the project may be calculated for each delivery option. This is shown at the bottom of Table 5.

The NPV estimates for the hypothetical project presented in Table 5 indicate that public costs are significantly higher for the two P3 options when the long-term risks to be retained under PSC are not considered. The decision on whether or not the P3 options provide VfM will rest on whether the cost increments are worth the transfer of long-term risks – for lifecycle performance only in the case of the AP concession, and for both lifecycle performance and revenue risk in the case of the toll concession option.

Summary and Conclusion

A case study review of seven VfM studies conducted in the United States suggests that VfM analysis approaches are inconsistent and their results are often confusing for a lay audience. It is often difficult to decipher the source of differences between delivery options by viewing VfM analysis results. Discount rates are sometimes used in a way that reduces credibility of the results, for example by producing "negative" financing costs. This paper has therefore attempted to develop a framework that may be used to ensure credible, consistent, and transparent presentation of VfM analysis results from a societal perspective. Project sponsors may consider elements relative to their own perspective while still using the framework.

Acknowledgments

The author acknowledges the valuable comments received on an earlier version of this paper by Marcel Ham of IMG Rebel and by TRB's Committee on Revenue and Finance.

Author Contributions

The author, PD-S, confirms sole contribution to the paper as follows: study conception and design; data collection; analysis and interpretation of results; and draft manuscript preparation. The author reviewed the results and approved the final version of the manuscript.

References

1. Government of Australia. *National Public-Private Policy Guidelines Volume 2: Practitioners' Guide*. Government of Australia, 2015.
2. HM Treasury. *Value for Money Assessment Guidance*, August 2004 (updated November 2006).
3. Partnerships British Columbia. *Methodology for Quantitative Procurement Options Analysis*, 2011 (updated April 2014).
4. Federal Highway Administration. *P3-Value 2.0 Guide*. U.S. Department of Transportation, Washington, D.C., February 2016.
5. Stuart, M. *Value for Money: Cautionary Lessons about P3s from British Columbia*. Canadian Centre for Policy Alternatives, BC Office, 2006.
6. Morallos, D., A. Amekudzi, C. L. Ross, and M. Meyer. Value for Money Analysis in U.S. Transportation Public–Private Partnerships. *Transportation Research Record: Journal of the Transportation Research Board*, 2009. 2115: pp. 27–36.
7. Federal Highway Administration. *Value for Money State of the Practice*. December 2011.
8. Rooney, M. *Technical Note: Lessons Learned Report*. March 21, 2012.
9. Barutha, P., and S. Eric. Value for Money on a Public Private Partnership Infrastructure Project: An International Comparative Analysis. Presented at 95th Annual Meeting of the Transportation Research Board, Washington, D.C., 2016.
10. National Infrastructure Unit of the Treasury. *Guidance for Public-Private Partnerships (P3s) in New Zealand*. NZ Treasury, 2009.
11. Gifford, J. L., P. K. Wheeler, and J. Y. Kweun. Comparative Analysis of Value-for-Money Studies: Highway Public-Private Partnerships Projects in the United States. Paper number 16-4546. Presented at 95th Annual Meeting of the Transportation Research Board, Washington, D.C., January 2016.
12. Boardman, A. E., and M. Hellowell. Comparison and Analysis of Specialist PPP Units' Methodologies for Conducting Value for Money Analysis. *Journal of Comparative Policy Analysis: Research and Practice*, pp. 1–17. DOI: 10.1080/13876988.2016.1190083
13. Colorado High Performance Transportation Enterprise. *US 36 Managed Lanes Toll Concession Project: Project Value Analysis*. Colorado High Performance Transportation Enterprise, Colorado, 2014.
14. Virginia Office of Public-Private Partnerships. *Transform 66 Project (Outside the Beltway): Initial Value for Money Assessment*. Virginia Office of Public-Private Partnerships, Virginia, 2015.
15. Colorado Department of Transportation. *I-70 East Project Financing and Delivery Options: A Summary of the Value for Money Analysis*. Colorado Department of Transportation, Colorado, 2015.
16. Florida Department of Transportation. *I-595 Corridor Roadway Improvements: Value for Money Analysis*. Florida Department of Transportation, Florida, 2009.
17. Ohio Department of Transportation and Kentucky Transportation Cabinet. *Brent Spence Bridge Project Options Analysis*. Ohio Department of Transportation and Kentucky Transportation Cabinet, Ohio, 2013.
18. Arup/PB Joint Venture. *Analysis of Delivery Options for the Presidio Parkway Project*. Arup/PB Joint Venture, San Francisco, 2010.
19. Ohio Department of Transportation. *SCI-823 Portsmouth Bypass: Project Delivery Alternatives Analysis*. Ohio Department of Transportation, Ohio, Undated.

The Standing Committee on Revenue and Finance (ABE10) peer-reviewed this paper (18-00641).

The views expressed in this paper are those of the author and not necessarily those of the US DOT or FHWA. As the sole author of this paper, I accept full responsibility for the entire work and approve the final version of the manuscript.

Article

TRR
JOURNAL OF THE TRANSPORTATION RESEARCH BOARD

Lessons Learned from 40 Years of Local Option Transportation Sales Taxes in California

Transportation Research Record
2018, Vol. 2672(4) 13–22
© National Academy of Sciences:
Transportation Research Board 2018
Article reuse guidelines:
sagepub.com/journals-permissions
DOI: 10.1177/0361198118782757
journals.sagepub.com/home/trr
SAGE

Jaimee Lederman[1], Anne Brown[1], Brian D. Taylor[1], and Martin Wachs[1]

Abstract

Jurisdictions across the United States have increasingly turned to local option sales taxes, or LOSTs, to fund transportation projects and programs. California is an enthusiastic adopter of these measures; since 1976, residents in over half of the state's 58 counties have voted on 76 LOST measures. As of 2017, 24 counties, home to 88% of the state's population, have LOST measures in place. Many counties have enacted multiple measures, with passage rates especially high among renewal and follow-on measures. This research is the first comprehensive analysis of LOST measures; drawing on measure expenditure plans to determine the range and frequency of transportation projects and services funded. This detailed review of expenditure plans across dozens of urban, suburban, and rural California counties offers insight on these measures and the projects and programs they fund. Overall, this study finds that LOSTs are heterogeneous, often including something for nearly every interest group. Almost all of the measures studied dedicate funding to a mix of transportation modes, including highways, public transit, local road maintenance, and active transportation. Expenditures on particular modes vary, reflecting transportation geography across counties. On average, 60% of LOST expenditures in California fund road projects and over 30% are allocated to public transit. Measures often dedicate a substantially larger share of revenue to transit relative to transit's mode share. Finally, LOSTs typically appeal to diverse local interests by returning a portion of revenues to local jurisdictions to address local priority projects.

Inflation-adjusted federal funding of transportation projects per vehicle mile of travel has been decreasing for decades. In response, states and regions around the United States have scrambled to make up the shortfalls. In the nation's most populous state, California, the 18-cent per gallon motor fuel excise tax remained unchanged between 1993 and 2016, despite the fact that inflation and increased vehicle fuel efficiency together eroded the fuel tax's buying power (*1*). Since the 1980s, the growing gap between transportation program needs and revenues has been increasingly backfilled in California (and several other states) by county sales tax measures for transportation.

Since 1976, California residents have voted on 76 local option sales taxes (LOSTs) to fund transportation in 30 of the most populous of the state's 58 counties. As of 2017, 24 counties, home to 88% of the state's population (*2*), have active LOST measures. The most recent measures in these so-called Self Help Counties have been approved by at least two-thirds voter majorities as required by California law. Transportation sales tax measures can produce substantial revenue for the maintenance, operation, and expansion of transportation facilities and services; sales tax revenues dedicated to transportation today produce over US$4 billion per year for transportation construction and maintenance in California (*3*). Voters in some counties have approved LOST measures as many as five separate times (*4*).

LOSTs for transportation have proven politically popular; voters tend to like them because they fund popular projects via taxes automatically levied in very small increments (often a ½- or 1-cent per dollar) over a very large number of transactions. Because these small levies apply to nearly all consumer purchases, LOSTs generate substantial revenue for transportation. On the other hand, LOSTs are less related to travel than are fuel taxes or tolls, which means that light users of transportation systems tend to pay more per mile in transportation sales than do heavy users (*5*).

Despite LOST's increasing role in transportation finance, there is no comprehensive, up-to-date data source on transportation sales tax measures. Given the labor intensiveness of gathering comparable data on LOST measures, and in particular on what they fund, this research focuses on the most

[1]Institute of Transportation Studies, UCLA, Los Angeles, CA

Corresponding Author:
Address correspondence to Jaimee Lederman: jaimeel@ucla.edu

populous state with the most LOST measures – California. Although both the Center for Transportation Excellence and the Eno Center for Transportation compile basic data on LOSTs and other local transportation funding measures nationwide, no previous research has assessed the categorical distribution of revenues based on expenditure plans, or placed measure success in the context of previous LOST measures put before voters in the same county (*6*, *7*). The data assembled for this research were used to conduct a thorough analysis of all 76 California LOST measures to produce the first comprehensive assessment of the breadth, diversity, and trends among LOST transportation measures within a single state. Because they focus solely on California sales taxes, results may not be representative of transportation sales taxes in other states, or non-sales-tax local transportation funding mechanisms, such as property taxes. Nevertheless, California's sheer size (12% of the US population) and nearly four decades of LOST experience offer important insights into this important and expanding form of transportation finance.

Policymakers at all levels of government are continually seeking ways to provide adequate, stable funding for transportation programs. Whether by increasing per gallon excise taxes—as California did in 2017 for the first time in nearly a quarter of a century—introducing tolls or high-occupancy/toll lanes, initiating new road use charges, or relying more heavily upon general revenues, including LOSTs, policymakers consider a broad spectrum of factors in deciding how to collect and expend revenues for transportation. The aim of this article, therefore, is to help transportation decision makers determine the outcomes of perhaps the most popular local transportation funding mechanisms—the sales tax.

Literature Review

The Political Appeal of LOSTs

Most studies of LOSTs focus on the factors influencing their success or failure at the ballot box. Previous work has found that many factors influence a measure's passage, including the development of expenditure plans (*8*, *9*), public marketing campaigns (*10*), and contextual and socioeconomic factors (*11*, *12*).

Public finance research has compared the revenue-raising potential of LOSTs in different types of jurisdictions. Multiple studies have found that LOSTs have the highest revenue potential where non-residents are likely to "import" money to the jurisdiction, including at retail centers (*13*) and where tourism is common (*14*). Findings are mixed with respect to their revenue generating capacity, and therefore attractiveness, in suburban and rural areas. Afonso finds that suburban counties in North Carolina (which are small and numerous) have relatively low revenue generating capacity, while Rogers finds that suburban municipalities in Oklahoma have high revenue potential compared with either cities or small towns in rural areas (*14*, *15*).

The Politics of Taxation

What Motivates People to Tax Themselves? Motivations to vote for tax increases have been explored extensively in political science literature. The influences and motivations for self-taxation generally fall into two categories relating to whether voters are affected by the specific tax measures' content (endogenous influence) or other factors (exogenous influence).

Exogenous influences on choices of self-taxation include the current economic climate, the manner by which services are currently paid (i.e., the status quo), the composition of current taxes, and the partisanship of the constituency. The importance of economic climate is reflected in findings that initiatives are less likely to pass during recessions than in expanding economies (*16*). Voters also appear to strongly consider their local jurisdiction's status quo in taxation: one study found people preferred taxes to user fees where similar goods or services were already paid for through taxes, and that they preferred fees to taxes where no related taxes were in place (*17*). Similarly, the composition of existing taxes matters: people "seem willing to consider higher total tax burdens if there are more smaller taxes" (*18*).

Endogenous influences on voters' willingness to tax themselves include the content of the initiative, spending by both the supporters and opponents of the tax, how tax impacts are perceived, how taxes are named or labeled, what the perceived benefits from the tax-funded spending are, and whether existing government services are viewed favorably. The amount of campaign money spent opposing a measure also increases the likelihood of failure, whereas spending in favor of a measure is less effective (*16*, *19*). For transportation sales taxes specifically, Manville and Cummins found that supporters are more likely to be motivated by collective benefits than private ones (*20*). For example, voters support transit-funding measures out of concern for public ills like congestion and pollution rather than a personal desire to use transit themselves; the authors note that a 2013 American Public Transportation Association study showed that 70% of Americans support increased transit spending even though relatively few people use transit, which accounts for fewer than 3% of all trips (*20*).

Literature that bridges political science and cognitive psychology finds that voters' willingness to approve taxes depends in part upon their perceptions of such taxes' impacts as well as the names given to the taxes. Perceptions of tax impacts tend to change when taxes are presented in percentage rather than dollar terms, and people support "both higher and more steeply progressive taxes" when they are stated as percentages (*18*). Finally, the perception of existing transportation service matters. A study of transportation tax measures in two southern California counties found that "respondents in both counties are approximately 7% more likely to support the extension of the transportation sales tax measures when [existing] transit options are viewed favorably." However, an unfavorable view of existing transit does not cause significant opposition (*21*).

Equity Issues in LOSTs

Goldman and Wachs note that sales taxes are inherently regressive (*13*). Despite this prevailing view of sales taxes in general, transportation sales taxes are often perceived as "fair" for several reasons. First, sales taxes encourage horizontal equity to the extent that groups with similar incomes have similar expenditure patterns. Second, sales taxes cannot be easily evaded and are paid both by residents and by non-residents, who contribute to transportation infrastructure wear and tear. Third, some argue that LOSTs are a more equitable source of transportation funding than the gasoline tax, as users of non-automobile modes also pay to directly fund transportation under the sales tax (*13*).

Studies that evaluate the equity implications of transportation LOSTs often focus on geographic equity as a crucial element of successful measures. In a case study of multiple measures in Sonoma County, Hannay and Wachs found that proximity to enumerated projects correlated with the percentage of "yes" votes on proposed measures, which supports the notion that geographic equity—ensuring that all areas in the county directly benefit from the listed projects—may contribute to measure passage (*12*). Haas et al. noted that evidence of geographic equity principles built into LOST measures can be seen in the increasing local revenue return provisions that seek to return revenues to localities in proportion to the revenue raised (*10*). However, the authors were unable to conclude whether or to what degree geographic equity affects the passage of LOSTs.

Finally, LOST equity may be evaluated on the basis of the modal mix of projects funded in a measure. After multiple failures, Hannay and Wachs partially attributed the eventual passage of Sonoma County's Measure M (2004) to its incorporation of a multi-modal funding plan (*12*). Haas et al. found limited evidence that a balance among highway and transit projects increased the likelihood of voter approval, which they attributed to the influence of interest groups, such as environmentalists, who prefer specific modal outcomes (*10*).

While informative, the studies cited above typically analyze one or just a few LOST measures, and often through qualitative case study approaches, which limit their generalizability. This research seeks to address this limitation and tests several of the hypotheses proffered in previous research through more comprehensive analysis of LOST measures.

Research Design and Methods

This study presents an overview and analysis of all 76 LOST measures put before California voters between 1976 and 2016. For each measure, the official text, published expenditure plans, county websites, and official Statements of Vote from county voter registrars were used to gather 21 unique variables spanning five general categories: administration, temporal factors (such as the period specified for the life of the tax), financial characteristics, voter support, and modal funding splits. The data gathered for all 76 California LOST transportation measures are available for download at www.its.ucla.edu/LOSTdata. The analysis presented here includes all measures for which data were available for a given variable.

Complete records were obtained for most measures and variables, but modal funding split data are limited to the 53 measures with available expenditure plans. In general, measures excluded from the modal split analysis are older and more likely to have failed compared with those included. Of the measures with expenditure plans, over two-thirds passed (68%), compared with just over half (52%) of measures without expenditure plans. Thus, the modal funding analyses offer an unavoidably incomplete picture of LOST expenditure plans. Nonetheless, the data presented here represent the most comprehensive analysis of California LOSTs to date.

For each of the 53 measures for which expenditure plans were available, expenditures were divided into six categories: public transit, bike/pedestrian facilities, seniors/disabled services, safe routes to school, local roads, and highways. Classification of expenditures by mode relied only on project lists and funding categories outlined in the measures' expenditure plans, which are published when a measure is put before voters. Transit funding reflects spending on all types of public transit, including rail and bus rapid transit capital investments as well as funding for transit operations and fare subsidies. Bike and pedestrian funding included a range of projects such as sidewalk repairs and new bike lanes. Seniors/disabled and safe routes to school refer to specific allocations to these categories in expenditure plans. Specific highway projects, as well as funding for "regional" road projects, are categorized as "highways." In cases in which the expenditure plan modal funding breakdown differed from a measure's list of projects, data from project lists were used because project lists provided more detailed enumeration of expenditures that are likely to be a more accurate representation of modal expenditures.

Local road funding was calculated from two sources. Local road funding figures are primarily based on the amount of funding specifically allocated to local roads in a measure's expenditure plan. If measures did not specify the percentage of funding allocated to local roads, local road funding was calculated as part or all of the funding dedicated to local return. For example, a measure may require that a minimum or specified percentage of the revenues be dedicated to specific uses (for example, local transit services, bike paths, and complete streets programs). However, in many other measures, local return funding is flexible, allowing local jurisdictions to decide which transportation projects to prioritize over the life of the measure. In these cases, any flexible local return funding not dedicated to a specific modal use was assumed to be local road funding, which evidence (detailed below) suggests is by far the most common use for local return funds. Though not ideal, this assumption reveals the dilemma in accounting for these funds. These funds are not specifically designated until the time of expenditure, which over time can be many years after the passage of the measure; even then, the local roads projects receiving LOST funding are often not publicly designated as LOST-funded projects, making the modal accounting very difficult. The choice was thus to either make some reasonable assumption about the expenditure of these funds, or exclude them from the analysis entirely, which would clearly distort the portrait of LOST funding. Given this, it was assumed that local return funds went to streets and roads expenditures for three reasons. First, local roads are the single most common expenditure category for local return funds when the modal splits are known (see, for example, Imperial County Measure D [*22*]; Riverside County Measure A2 [*23*]). Second, measure expenditure plans typically describe flexible funding as local road funding unless there is a portion dedicated to a specific mode (for example Ventura County Measure AA, 2016). Third, ballot arguments overwhelmingly tout the benefits of local return funding for local road repair and maintenance (*24*).

Many measures include projects or expenditures that either do not fit into any of these modal classifications, or span multiple modes. For example, measures often earmark funds for regional planning or environmental mitigation. Some funds may also be double-counted across categories for legitimate reasons. For example, if a percentage of revenues is dedicated to fare subsidies for seniors and disabled riders, it was counted as both "transit" and "seniors/disabled." Additionally, local return funding is flexible and in many cases can be used for alternative transportation or other planning exercises (for example, complete streets planning). Thus, the modal breakdown likely underestimates the actual share of measure revenues spent on non-automobile modes. Lastly, a number of measures dedicate funding to competitive grant programs that are not limited to a particular mode. This funding cannot be categorized at the time of the election and is therefore not considered in the modal breakdown. For these reasons, modal funding categories may not sum to 100%.

Modal breakdowns of expenditures were used to calculate how closely a measure's expenditure plan reflects the current commute travel modes of county residents. Commute mode share was used for comparison rather than share of all trips because work trip data are typically included in regional transportation plans and frequently cited by measure proponents and opponents. Using countywide commute share from the 2015 5-year American Community Survey and previously calculated modal expenditure breakdowns, modal balance was calculated for mode *m* using the following formula for each county:

$$\textit{balance}\ (m) = \textit{percentage share of expenditure plan}\ (m) \\ -\textit{countywide commute share}\ (m)$$

Modal balance was calculated for only two modes: highways/local roads and transit. Walking and biking infrastructure was omitted because, though they are an important and growing part of mobility, the commute mode share data are classified with "other" modes in American Community Survey data, making determination of commute share by the combined category of walking and biking impossible. Positive balances indicate that a measure dedicated a greater share of funding to a mode relative to the share of commuters taking that mode. A negative balance indicates that a measure dedicated a lower share of revenues relative to the share of commuters taking that mode. "Overall" balance is then calculated as the sum of the absolute values of the transit and car balances. Higher numbers indicate *less* overall measure balance.

Findings

The first California LOST was enacted in Santa Clara County in 1976; since then, LOST measures have become increasingly popular means for counties to finance their transportation systems (See Figure 1). More popular than ever, over half (39) of all LOST measures were proposed in the last decade (2006–2016), despite the Great Recession's depressing effects on ballot measures between 2008 and 2014.

Figure 1 shows the number of measures proposed each year, distinguishing between "newcomer" counties (those proposing their first LOST measure), and "veteran" counties that had previously put a LOST measure on the ballot. Before 2000, 15 California counties had proposed LOST measures. Between 2000 and 2016, an additional 17 counties proposed LOST measures, with four new counties doing so in 2016 alone.

Measure Passage Rates over Time

In 1995, the California Supreme Court ruled in *Santa Clara County Local Transportation Authority v. Guardino* that, since LOST revenue was required to be spent on transportation, LOSTs are considered a "special" tax for a specific purpose (as opposed to a "general" tax). Under California's Proposition 62, such special taxes require approval by a two-thirds supermajority of voters (*25*). Prior to this decision, only a simple majority was required to approve a LOST, but since 1995 each "no" vote on a LOST measure effectively counts twice as much as a "yes" vote. Following *Guardino*, some feared that the era of sales tax transportation finance would be short-lived. Indeed, six of the nine measures (67%) placed on the ballot between 1996 and 2002 following the *Guardino* ruling received a majority of the votes cast in favor of passage, but failed to clear the new two-thirds supermajority hurdle. Since those few tumultuous post-*Guardino* years, however, the average share of "yes" votes for these measures has climbed, as has the rate of passage under the supermajority requirement. Between 2003 and 2016, no fewer than 50 LOST measures were put on the ballot:

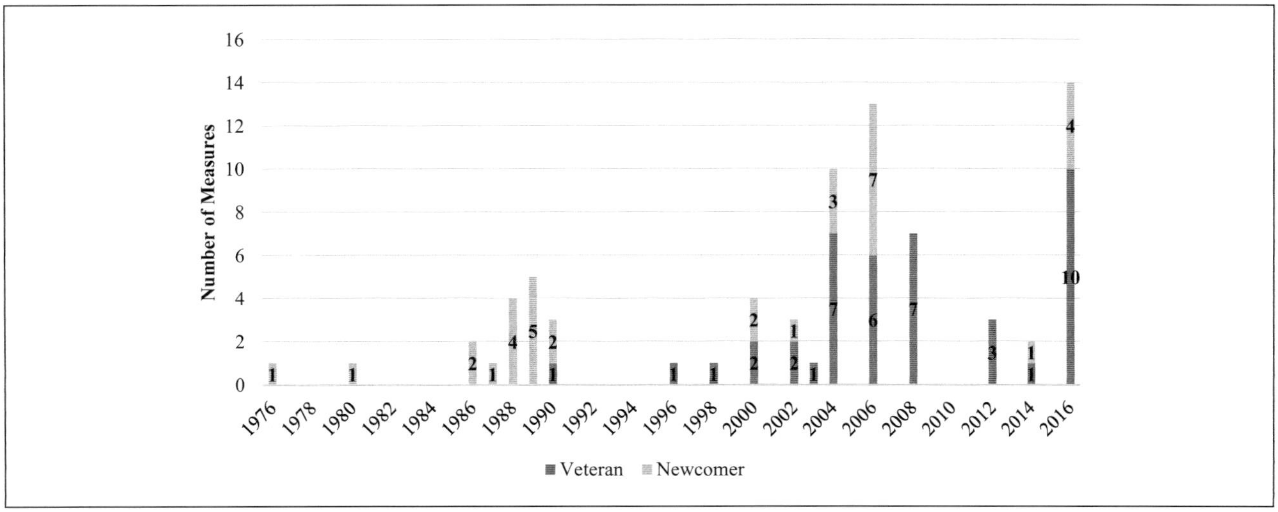

Figure 1. Veteran and newcomer county measures by year.

- 8 of the 50 (16%) failed even to garner a simple majority of the votes cast;
- 15 of the 50 (30%) received a simple majority of votes in favor, but not the required supermajority; and
- 27 of the 50 (54%) received supermajority support and were enacted.

Figure 2 shows the percentage of voter approval for every measure in the sample, as well as the passage thresholds needed under the earlier simple majority and later supermajority rule; Overall, 48 of the 76 measures (63%) passed, whereas 28 of the 76 (37%) were rejected—eight in 2016 alone. Figure 2 clearly demonstrates the significant implications the majority rule change had on passage rates. All 17 measures proposed before the supermajority rule were approved, whereas about half of measures (53%) proposed following the supermajority requirement passed. Of the 28 failed measures, all but eight would have been enacted under a simple majority requirement. The divergence between what would have been approved under the two different majority requirements underscores the importance of the supermajority ruling on LOST outcomes in California.

Measure Characteristics

Measure Type and Tenure. More than half of the 76 (57%) LOST measures appeared on ballots in counties that had not previously enacted a transportation sales tax. An additional 21 measures (28%) extended or renewed LOST measures already in place. Only five (7%) measures are permanent; four of these passed (three in Los Angeles County and one in Santa Clara County). The clear majority (71 of 76, or 93%) of California LOSTs are termed measures, meaning the increased sales tax levy is in place for a fixed period, after which it expires. It is therefore unsurprising that renewals and extensions have become increasingly common as original sales tax measures reach their sunset dates.

The durations of termed measures range from 8 to 40 years, with an average of 24 years; more than three-quarters of LOSTs are slated to last at least 20 years. The shortest proposed LOST period was Sonoma County's unsuccessful 2000 Measure B, which proposed a 0.5% sales tax for 8 years.

Renewal and extension measures were more likely to pass (58%) than were original sales tax measures (53%), even though all extensions and renewals were subject to the supermajority requirement whereas only two-thirds (69%) of original measures were required to pass with a supermajority. Among measures subject to the supermajority requirement, passage rates differ between original measures and renewals/extensions. Only 30% of original measures following 1995 were approved, while 72% of additional measures in counties that had previously passed measures were successful in the same period.

This suggests that as counties demonstrate their ability to deliver projects from past measures, voters are more willing to vote for subsequent measures. A second possible explanation is that the temporal distribution of original measures—those proposed by counties without previously passed LOSTs—represents a political, geographical, or cultural divide between urban and more rural counties. For example, since the supermajority requirement, many rural or conservative counties have placed a LOST on the ballot for the first time (*22*). Residents of mostly urban counties may simply be more supportive of LOST measures, transportation spending, or taxation more broadly, compared with residents of more rural counties. Future research is needed to better understand factors influencing measure support across geography and populations.

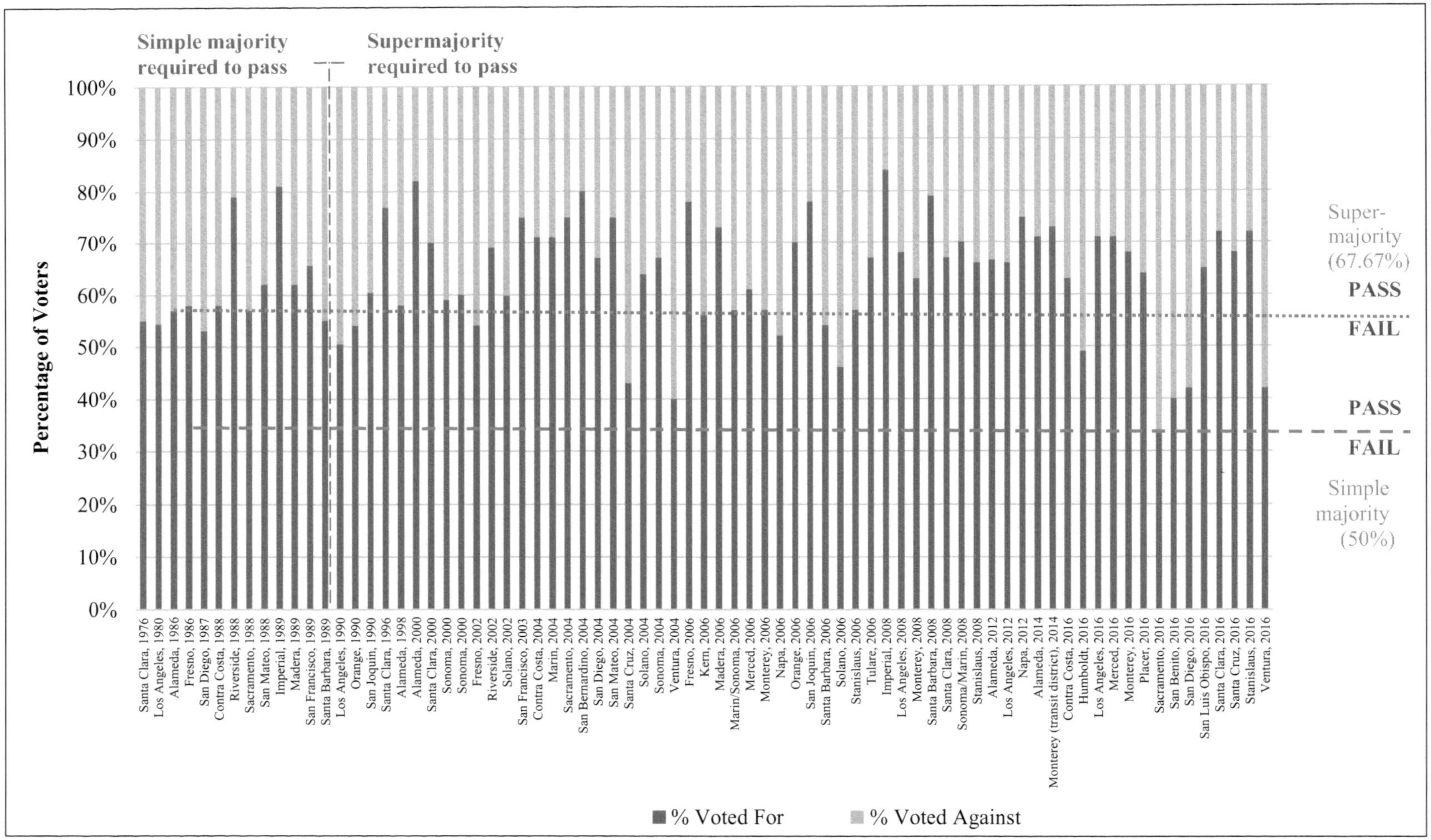

Figure 2. Measure of support over time.

Tax Rates and Expected Revenues. The proposed sales tax increases range from 0.125% (Monterey, 2014; Santa Clara, 2008) to 1% (Alameda Measure B, 2012; Alameda Measure BB, 2014), although a substantial majority (86%) of LOSTs impose a half-percent (alternatively referred to as a half-cent) sales tax. The small rates levied in the 2014 Monterey measure (Measure Q) and 2008 Santa Clara measure (Measure B) fund specific public transit improvements, narrowly defined projects, or projects that are both public transit improvements and narrowly defined. The half-cent sales tax rate has remained popular over time; a half-cent is the median sales tax increase amount in every election year, except for 2014.

LOST expenditure plans forecast a wide range of annual revenues from the tax increases, reflecting counties that vary widely in both population and economic composition. Expected annual revenues ranged from a low of US$2.5 million (Monterey Measure A, 2006) to US$1.3 billion (Los Angeles Measure R, 2008), with an average of US$183 million per year. As one might expect, county population is strongly and positively correlated (0.44) with expected annual revenue from these measures.

Expenditure Plan Characteristics

Modal Funding Split. Table 1 shows the large range in funding of various transportation modes across the measures, drawn from the 53 measure expenditure plans. Although most measures split funding among modes, a few proposed to dedicate all tax revenues to a single recipient. For example, Monterey's 2008 Measure Q and Santa Clara's 2000 Measure A and 2008 Measure B dedicated 100% of revenue to public transit projects, whereas Humboldt County's 2016 Measure U proposed to return all funding to local jurisdictions.

Each funding category developed for this analysis includes a wide variety of investment types. Qualitative analyses of the expenditure plans reveal that most itemized projects are capital investments rather than operating or maintenance expenditures.

Most measures contained at least some funding for transit (91% of measures), local roads (87%), and highways (83%). Although most counties allocated some funding for transit and other non-automobile modes, the combined funding for local roads and highways accounted for an average of 61% (and a median of 69%) of funding across all measures. This is far higher than spending for public transit, which received an average of 31% (and a median of 26%) of dedicated funding across the measures. Over half (55%) of the measures allotted at least some funding to bike or pedestrian projects, or to both bike and pedestrian projects.

Modal Funding Balance. The programs detailed in these measures reflect both the diversity of counties across California, and each county's unique dynamic of achieving of local political consensus. Counties exhibit different approaches to balancing current transportation needs on one hand against aspirations for shifting future travel behavior and transportation modes on the other. One way to cast light on this "needs" versus "aspirations" dichotomy is to compare the expenditures proposed in the plan with current travel patterns. No LOST expenditure plan precisely mirrors the commute mode share in a county. In general, the plans tend to fund public transit at much higher levels than the share of people who ride transit to work, the trip purpose that enjoys the highest transit mode share. Figure 3 shows measures ranked from most to least balanced between expenditures and commute mode. Each measure's overall balance is listed in parentheses next to the measure name.

Figure 3 shows that, among the 53 measures that included expenditure plans, all but seven (87%) allocated a higher share of measure expenditures to public transit than the public transit commuting mode share in that county. Two county measures (Alameda 1986, Santa Cruz 2004) proposed public transit expenditures at levels roughly equivalent to the transit commute mode share. The remaining five counties dedicated either most (Imperial 1989, 95%; Napa 2012, 92%; Stanislaus 2008, 98%) or all (Humboldt 2016; San Benito 2016) funding to streets and highways, proportionally higher than motor vehicle travel share.

Although the majority of expenditures across all of the county measures (61%) is allocated to streets and highways, more than four out of five (83%) commuters travel by car in the California counties with LOST expenditure plans (*2*). Collectively, these data suggest three possibilities. First, the authors and supporters of county expenditure plans may aspire to support higher levels of public transit use and less dependence on driving. Second, proportionately higher public transit funding in the LOST measures might reflect the current structure of transportation funding in California,

Table 1. Modal Funding Split

Mode	Mean	Median	Range	25th–75th percentile
Highways	27.0%	26.0%	0–66%	9–44%
Public transit	31.0%	26.0%	0–100%	10–43%
Local roads	34.0%	30.0%	0–100%	20–40%
Bike/pedestrian	2.4%	0.7%	0–17%	1–4%
Safe routes to school	1.8%	0.0%	0–11%	0–0%
Seniors/disabled	4.4%	1.0%	0–100%	0–4%

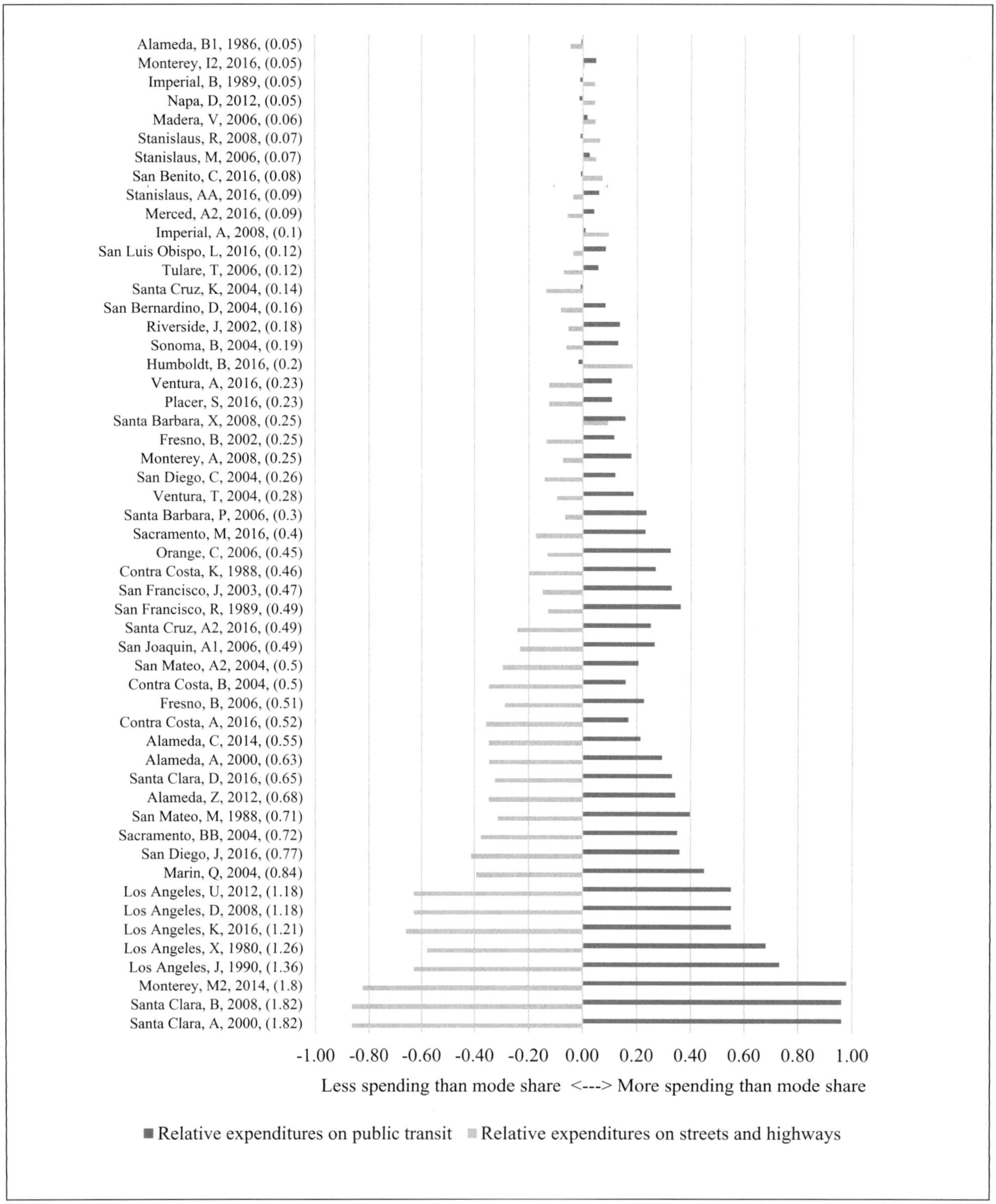

Figure 3. LOST measure balance and proportional funding.

whereby streets and highways receive more funding from traditional revenue sources such as the fuel and property taxes than does transit (*26*). Or third, transit funding may be a political calculus whereby the measures' authors may perceive transit to be relatively popular with voters and conclude that LOST transit expenditures will buoy the odds of passage even if many voters do not personally intend to ride transit. Hannay and Wachs and Haas et al. report that measures are more likely to pass if they dedicate funding to a mix of highway and transit projects (*10*, *12*). However, although the majority (85%) of measures *do* include both highway and transit funding, the simple inclusion of both was not enough to increase a measure's chances of passing; in fact, it did not meaningfully explain whether a measure passed or failed.

Addressing Geographic Equity Concerns. Historically, parts of counties slated to receive fewer transportation benefits than they contribute in sales tax revenue are less likely to vote for a LOST transportation measure. This could easily lead to defeat in California given the supermajority requirement (*10*, *12*). Measure authors appear to pay careful attention to addressing geographic equity in two ways. First, measure proponents typically assemble project lists to fund major capital projects in all parts of the county. Second, almost all measures dedicate a portion of revenue for return directly to cities (and the county for unincorporated land) as flexible funding for local priority projects. In most cases, the clear majority of local return is spent on local road projects (as discussed above). Some measures stipulate that a certain portion of local return be dedicated to specific modes (often bike and pedestrian infrastructure), although jurisdictions are free to select specific projects.

Local return was first featured in Los Angeles County Measure A in 1980, and average local return percentages have increased over time. For the 49 measures that included local return funding in expenditure plans, an average of 35% of measure revenues were dedicated to local return.

Figure 4 shows the distribution of local return expenditures among the studied measures. On average, measures returned about one-third (34%) of revenue to local jurisdictions. Each county has unique transportation geography, which LOST measures tend to reflect. Several measures have local return values that appear to be outliers, but which logically reflect local geography. For example, revenues from Santa Clara Measure A (2000) and Measure B (2008), and Monterey County Measure Q (2014), were entirely dedicated to specific transit projects, and thus have zero percent local return funding. At the other extreme, four measures (in relatively low population counties where big projects are rare) dedicated over 90% of measure revenues to local return: Imperial County Measure D (1989) and Measure D (2008), Napa County Measure T (2012), and Humboldt's Measure U (2016).

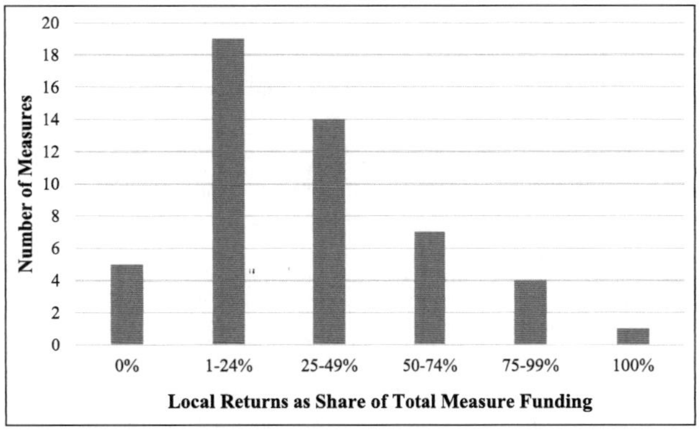

Figure 4. Distribution of local return as share of measure revenues among measures studied.

Conclusion

Although conventional wisdom holds that voters do not like to tax themselves, LOST measures have proven to be remarkably popular with voters over the years. In California, nearly two-thirds of LOST measures put before voters have been approved—most by a supermajority—and their popularity as a transportation finance tool appears to be waxing. Fourteen measures were on the 2016 ballot alone, four in counties that had never before considered such measures. In addition, the rate of LOST measures' passage has increased over time, making LOSTs politically effective mechanisms for garnering popular support to raise taxes for transportation.

Each of the LOST expenditure plans examined here was distinct in some way, reflecting each county's unique urban geography and transportation needs. Most measures dedicate the largest portion of expenditures to road projects, followed by public transit projects – with suburban and rural counties earmarking a higher share of revenue for roads, and urban counties devoting larger shares to public transit. On average, about three-fifths of all LOST expenditures go to road projects (34% for local roads, and 27% for highways, on average), 31% is allocated to public transit projects, and about 8% is dedicated to benefiting specific groups such as projects for the elderly and disabled, bicycle and pedestrian facilities, and safe routes to school programs. Most measures also include a share of funds slated for "local return," which spreads non-mode-specific funding around the jurisdiction by reserving a portion of the funds to be used at local discretion.

Although modal funding expenditures vary widely among counties, they rarely reflect local mode shares. Most often, LOST measures allocate a percentage of funds to public transit that is proportionately greater than transit usage. This may reflect policy aspirations for transit as an alternative to auto dependence; it may reflect that public transit has fewer revenue sources to draw on than streets and highways (*26*); or it may reflect a multi-modal political strategy to secure a broad base of popular support (*10*, *12*), or a combination of these factors. Manville and Cummins found that people often support transit spending based on a belief in its collective benefits—such as improved air quality and reduced congestion—rather than a personal intention to ride transit and derive a private benefit from doing so (*20*).

The LOST measures examined here have detailed expenditure plans, are mostly multi-modal, often provide substantial public transit funding together with flexible local return funding, and most have fixed-terms requiring voter-approved renewal. Together, these characteristics have combined to create a winning transportation finance formula resulting in enactment of many measures in California, and around the United States, despite (in the California case) a supermajority voter approval requirement.

Acknowledgments

The authors thank the many parties who made this project possible. First, we thank the University of California Center for Economic Competitiveness (UCCONNECT) who generously funded this project.

Author Contributions

The authors confirm contribution to the paper as follows: study conception and design: JL, AB, BT, MW; data collection: JL, AB; analysis and interpretation of results: JL, AB; draft manuscript preparation: JL, AB, BT, MW. All authors reviewed the results and approved the final version of the manuscript.

References

1. Perry, F. N., C. Kredell, M. E. Perry, and S. Leonard. Beyond the Gas Tax: Funding California Transportation in the 21st Century, 2017. http://www.next10.org/sites/next10.org/files/transportation-funding-brief-final.pdf.
2. U.S. Census Bureau. 2011-2015 5-Year American Community Survey. 2015.
3. Wachs, M. The Future of Transportation Finance: A New Generation of User Fees? *Weekly Policy Commentary, Resources for the Future*, Vol. 14, 2010. http://www.rff.org/blog/2010/future-transportation-finance-new-generation-user-fees.
4. Self-Help Counties Coalition, 2018. http://selfhelpcounties.org.
5. Dill, J., T. M. Goldman, and M. Wachs. California Vehicle License Fees: Incidence and Equity. *Journal of Transportation and Statistics*, Vol. 2, No. 2, 1999, pp. 133–147.
6. Center for Transportation Excellence. *Transportation Ballot Measures*, 2018. http://www.cfte.org/elections/past.
7. Eno Center for Transportation. *The Life and Death of the Highway Trust Fund*. Eno Center for Transportation, 2014. https://www.enotrans.org/wp-content/uploads/2015/09/Highway-Trust-Fund.pdf.
8. Beale, H., E. Bishop, and W. Marley. How to Pass Local Option Taxes to Finance Transportation Projects. *Transportation Research Record: Journal of the Transportation Research Board*, 1996. 1558: 74–82.
9. Crabbe, A. E., R. Hiatt, S. D. Poliwka, and M. Wachs. Local Transportation Sales Taxes: California's Experiment in Transportation Finance. *Public Budgeting & Finance*, Vol. 25, No. 3, 2005, pp. 91–121.
10. Haas, P. J., K. S. Massey, L. O. Valenty, and R. Werbel. *Why Campaigns for Local Transportation Initiatives Succeed or Fail: An Analysis of Four Communities and National Data*, Mineta Transportation Institute, Report 00-1, 2000. http://transweb.sjsu.edu/MTIportal/research/publications/documents/00-01.pdf.
11. Hamideh, A., J. E. Oh, S. Labi, and F. Mannering. Public Acceptance of Local Government Transportation Sales Taxes: A Statistical Assessment. *State & Local Government Review*, Vol. 40, No. 3, 2008, pp. 150–159.
12. Hannay, R., and M. Wachs. Factors Influencing Support for Local Transportation Sales Tax Measures. *Transportation*, Vol. 34, No. 1, 2007, pp. 17–35.
13. Goldman, T., and M. Wachs. *A Quiet Revolution in Transportation Finance: The Rise of Local Option Transportation Taxes*. University of California Transportation Center, 2003.
14. Afonso, W. The Equity of Local Sales Tax Distributions in Urban, Suburban, Rural, and Tourism Rich Counties in North Carolina. *Public Finance Review*, Vol. 44, No. 6, 2016, pp. 691–721.
15. Rogers, C. L. Local Option Sales Tax (LOST) Policy on the Urban Fringe. *Journal of Regional Analysis and Policy*, Vol. 34, No. 1, 2004, pp. 25–50.
16. Donovan, T., and S. Bowler. *Demanding Choices: Opinion, Voting, and Direct Democracy*. University of Michigan Press, 1998.
17. Franko, W., C. J. Tolbert, and C. Witko. Inequality, Self-Interest, and Public Support for "Robin Hood" Tax Policies. *Political Research Quarterly*, Vol. 66, No. 4, 2013, pp. 923–937.
18. McCaffery, E. J., and J. Baron. Thinking About Tax. *Psychology, Public Policy, and Law*, Vol. 12, No. 1, 2006, p. 106.
19. Gerber, E. R. *The Populist Paradox: Interest Group Influence and the Promise of Direct Legislation*. Princeton University Press, 1999.
20. Manville, M., and B. Cummins. Why Do Voters Support Public Transportation? Public Choices and Private Behavior. *Transportation*, Vol. 42, No. 2, 2015, pp. 303–332.
21. Green, A. D., M. Neiman, S. Bockman, and B. Sirotnik. Public Support for Transportation Sales Taxes in California: A Two County Assessment. *California Journal of Politics and Policy*, Vol. 5, No. 4, 2013, pp. 645–670.
22. Imperial County. Measure D. 2011. http://www.selfhelpcounties.org/countyupdates/Imperial.pdf.
23. Riverside County. Measure A2 Ordinance No. 02-001. 2002.
24. Albrecht, M., A. Brown, J. Lederman, B. D. Taylor, and M. Wachs. The Equity Challenges and Outcomes of California County Transportation Sales Taxes. *University of California Center on Economic Competitiveness*, 2017. http://www.its.ucla.edu/publication/equity-challenges-outcomes-california-county-transportation-sales-taxes/.
25. Santa Clara County Local Transportation Authority v. Guardino. In *SCOCAL*, Cal. 4th, 1995.
26. California Department of Transportation. *Chart C: Budgetary Flow of California State Fees and Taxes Designated for Transportation Purposes Proposed for the 2014-2015 Fiscal Year*. California Department of Transportation, 2014.

The Standing Committee on Revenue and Finance (ABE10) peer-reviewed this paper (18-03384).

Article

TRR
JOURNAL OF THE TRANSPORTATION RESEARCH BOARD

Transportation Research Record
2018, Vol. 2672(4) 23–31
© National Academy of Sciences:
Transportation Research Board 2018
Article reuse guidelines:
sagepub.com/journals-permissions
DOI: 10.1177/0361198118794714
journals.sagepub.com/home/trr
SAGE

Estimating Revenue-Neutral Mileage-Based Fees for Urban and Rural Households in Eight Western States

Stephen S. Fitzroy[1] and Kyle Schroeckenthaler[1]

Abstract

This paper focuses on one alternative source of revenue—a mileage-based user fee—as a potential substitute for the current gasoline excise tax. It examines one of many questions that arise when this source of funding is considered—the potential for differential financial effects on urban and rural households. The issue has been characterized as one of "equity," and more specifically whether rural households will pay a disproportionately greater share of the costs if states transition from a fuel-consumption-based to a mileage-based funding system. This paper presents an approach to address this question using existing sources of data and a set of assumptions designed to assess a "revenue-neutral" substitution of mileage-based charges for the current gasoline excise tax. Based on the methods used in this paper, it appears that rural households may, in fact, benefit from introduction of a mileage-based fee because they will pay less than under the current system in all states investigated for this study.

Current limitations associated with reliance on the gas tax require consideration of alternative methods of generating revenue (*1*). The limitations and potential alternative sources of revenue generation have been discussed at length in prior studies (*2*). This paper focuses on one alternative source of revenue—a mileage-based user fee—and one of many questions that arise when this source of funding is considered—the potential for differential financial effects on urban and rural households. The issue has been characterized as one of "equity," and more specifically whether rural households will pay a disproportionately greater share of the costs if states transition from a fuel-consumption-based to a mileage-based funding system. This paper presents an approach to address this question using existing sources of data and a set of assumptions designed to assess a "revenue-neutral" substitution of mileage-based charges for the current gasoline excise tax. The estimation methods are designed to be reproducible and adaptable to other states or multistate regions interested in assessing urban/rural equity considerations.

Scope of Analysis and Assumptions

The paper is based on research conducted for eight of the 14 states currently members of the Western Road Usage Charge Consortium (RUC West). The states included in the analysis were:

1. Arizona
2. California
3. Idaho
4. Montana
5. Oregon
6. Texas
7. Utah
8. Washington

This research was designed to help determine both the direction and the order of magnitude of charges that households would be likely to bear if a state implemented a mileage-based charging system. Households were classified by their location as rural, urban or "mixed." In many areas of the Western U.S.—especially in areas with small and medium-sized population centers, the term "suburban" does not strictly apply. Households with a commute characterized by travel from a less dense location to a denser location fall into this "mixed" category.

The substitution of a mileage-based fee for a fuel-consumption-based tax was predicated on assumptions of "revenue-neutrality" with regard to current gas excise

[1]EDR Group, Boston, MA

Corresponding Author:
Address correspondence to Stephen S. Fitzroy: sfitzroy@edrgroup.com

taxes levied by each state. The approach did not rely on any supplemental surveys, interviews, or data collection. Rather, it was designed to be readily implementable for any state and updatable as conditions change. Other key assumptions included the following:

1. Census tracts were used as the geographic unit of analysis.
2. Only non-commercial vehicles registered at a place of residence were considered.
3. Data from publicly-available sources, such as the U.S. Census, were utilized as practical.
4. Non-gas-powered vehicles were included in the analysis to assess the mileage-based charges.
5. Diesel passenger vehicles would be considered in a separate study of diesel fuel taxes.

Literature Review

McMullen et al. (*3*) and Larsen et al. (*4*) review and synthesize the results of a decade of investigation on the urban/rural equity question based on setting flat vehicle miles traveled (VMT) fees and several variations. McMullen et al. (*3*) estimate the effects of a revenue-neutral mileage-based fee using Oregon Household Activity Survey (OHAS) data and a static estimation technique, concluding that under road usage charges (RUCs), rural areas would generally pay less than under the current excise tax system and that these differences would vary depending on fuel efficiency, as they do in urban areas. That research divided Oregon into eight regions based on the OHAS data. In addition to estimating RUC charges for Oregon (*3*, *5*) other statewide studies have included Nevada (*6*), Texas (*4*), and Iowa (*7*).

Weatherford focuses on a nationwide assessment of income and geographic equity of a uniform RUC using information for the 2001 National Highway Travel Survey (NHTS) and assuming a fixed mileage-based fee (0.98 cents per mile) (*8*). He concludes that the burden on rural and low-income households would be less regressive than the current national excise tax but also indicates that a rate set at this level would generally increase the tax burden for residents in the Pacific states. Other issues identified in the literature include the need to develop definitions of geographic areas that can provide detail not typically available in national data sets like the NHTS (*7*) and that provide more nuanced information about distinctions between urban and rural areas than are typically available from the U.S. Census Bureau (*9*).

A combination of demographics, geographic characteristics and vehicle fuel consumption attributes need to be considered when assessing the impacts of a RUC relative to the current gas excise tax. The urban/rural definition of the geographic region as well as the relative size of the region is also important. Moreover, it is also clear that, to date, there has been no methodology that distinguishes between revenue-neutral RUC rates for urban and rural areas and has been applied consistently across multiple states. This study was designed to address each of these issues and to provide practitioners with a uniform, implementable approach to estimating revenue-neutral RUC rates without having to resort to specialized surveys or complex econometric methods.

Approach

The approach to estimation of revenue-neutral mileage-based user fees required several steps (see Figure 1). Three basic types of information were required: a systematic urban/rural categorization system that could be applied to diverse states; the number of vehicle miles of travel associated with households in a census tract; and the mix of fuel types and efficiencies in each census tract. These three types of information were combined with gas tax rates for each of the states to estimate the gas taxes paid per household associated with the current fleet of gas-burning vehicles.

The objective was to estimate a mileage-based fee that would be equal to the portion of total gas tax revenues currently paid by households in each state. These "revenue-neutral" mileage-based fees were based on current estimated household gas excise tax payments, not total statewide revenue from all gas-related sources (e.g., sales or other taxes and fees, or excise taxes paid by commercial, agricultural, or recreational vehicles). Total gas-tax-based revenues were estimated using household VMT by census tract and the average fuel efficiency of gas-powered and flex-fueled non-commercial vehicle types registered in each census tract. The equivalent total revenue was then computed from mileage-based fees by including mileage for non-gas-powered vehicles and flex-fueled vehicles when computing a mileage-based fee.

Identifying Urban, Rural, and Mixed Households

There are many possible approaches to identifying urban and rural areas, but the one judged most suitable for this study was developed by the U.S. Department of Agriculture's (USDA) Economic Research Service (ERS). The ERS used census tract–level household data from the 2006 to 2010 American Community Survey (ACS) to identify 10 different types of commuting patterns within and between urbanized areas and less dense areas (Table 1).

Each census tract in the eight-state region has a Rural–Urban Commuting Area (RUCA) code assigned by ERS based on commuting patterns reported in the

Table 1. Rural–Urban Commuting Area (RUCA) Codes Developed by USDA

Code	Description
1	Metropolitan area core: primary flow within an urbanized area (UA)
2	Metropolitan area high commuting: primary flow 30% or more to a UA
3	Metropolitan area low commuting: primary flow 10% to 30% to a UA
4	Micropolitan area core: primary flow within an urban cluster (UC) of 10,000 to 49,999 (large UC)
5	Micropolitan high commuting: primary flow 30% or more to a large UC
6	Micropolitan low commuting: primary flow 10% to 30% to a large UC
7	Small town core: primary flow within an urban cluster of 2,500 to 9,999 (small UC)
8	Small town high commuting: primary flow 30% or more to a small UC
9	Small town low commuting: primary flow 10% to 30% to a small UC
10	Rural areas: primary flow to a tract outside a UA or UC

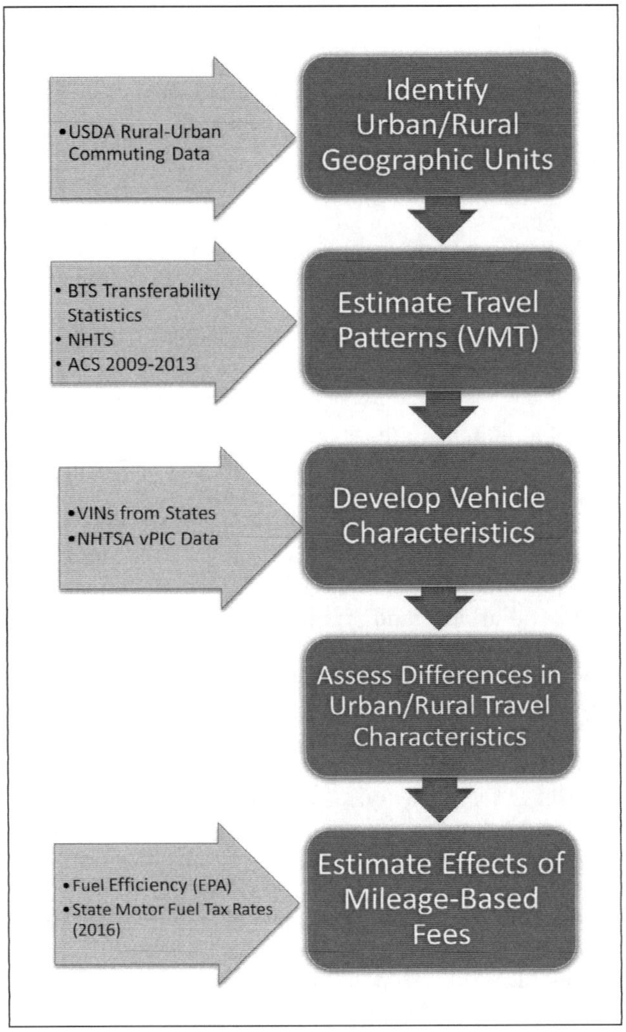

Figure 1. Study flow diagram.

ACS. Various combinations of these codes were evaluated for use in urban/rural classification. Table 2 shows the combination of RUCA codes that best captured urban and rural census tract commuting patterns for this study. A "mixed" classification was defined based on three RUCA codes that indicated primary flows to large urban areas (UAs) or strong flows to a large urban cluster (UC). The term "mixed" is used because some of these tracts have densities of less than 25 people per square mile on average in the western states. Mixed tracts are characterized by primary commuting flows to UAs; they contain less dense suburbs and outer suburbs and even stretch out from population centers to include land with more rural settlement and land use patterns. Socioeconomic and travel characteristics for these areas are different than for rural areas. Small towns that are dense enough to classify as UAs under the Census definition are often included in the "urban" category, not the "mixed" category for this study. The spatial patterns that emerged for the eight states are shown in Figure 2.

This method of classification picked up subtleties in commuting patterns between census tracts identified as smaller UCs and nearby rural census tracts. Urban census tracts include those in a metropolitan or micropolitan core with primary commuting flows between place of residence and work within the core area. Mixed census tracts consist of those in noncore metropolitan areas with primary flows to the metropolitan core, and census tracts with high levels of commuting between those located in noncore micropolitan areas and the micropolitan core. Rural census tracts are those census tracts with commuting flows within or between small UCs (populations under 10,000) and rural areas. Some urban census tract classifications may seem unusual considering their location in less populated counties; they generally have small UCs with most commuting internal to the census tract.

Estimating VMT by Fuel Type for Census Tracts

VMT by fuel type was approximated by estimating VMT by census tract and then using vehicle registration data to estimate the mix of fuel types at the census tract level.

Table 2. Aggregated RUCA Codes Defining Urban–Mixed–Rural Census Tracts

Urban–mixed–rural classification	RUCA codes included	Count of census tracts	Total land area (mi^2)	Total population
Urban	1, 4	12203	83,595	56,628,426
Mixed	2, 3, 5	1122	187,366	4,776,604
Rural	6, 7, 8, 9, 10	926	574,841	3,146,635

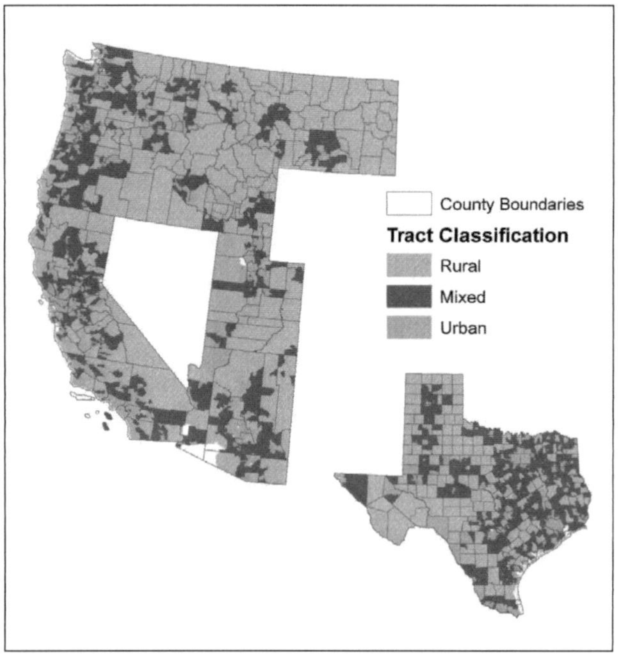

Figure 2. Urban/mixed/rural classifications by census tract.

Estimates of VMT are totals for households located in each census tract (not VMT for all vehicles passing through each census tract, regardless of where they are registered). Motor vehicle registrations from each state were matched with vehicle characteristics attributed using vehicle identification numbers (VINs) from registration data. Vehicle location was determined by matching registration data to census tract boundaries. Vehicle registration data was filtered to exclude commercial vehicles, farm vehicles, and other vehicles not registered to residential users (e.g., golf carts).

Estimates of VMT by fuel type were important for two reasons. First, estimates of VMT by gas-powered vehicles were necessary to determine the current level of gas tax being paid. This was accomplished by estimating average fuel efficiency for gas-powered vehicles for each census tract. VMT estimates were based on parameters developed by the Bureau of Transportation Statistics' (BTS) Transferability Statistics report from the NHTS (*10*). Second, a revenue-neutral mileage-based charge would then need to reallocate gas tax revenues paid by gas-powered vehicles to the total VMT estimated for both gas- and non-gas-powered vehicles.

The data needed to estimate VMT include independent variables available at the census tract level. These include the following:

- Average household income
- Average number of household vehicles
- Average number of household members
- Average number of workers
- Percentage of households who are homeowners
- Percentage of households with children
- Percentage of single member households
- Percentage of multiple member households, no members over 65
- Percentage of multiple member households, at least one member over 65

BTS parameters are estimated based on states aggregated by census divisions, and by urban, suburban, and rural areas within these divisions. Using the Urbanicity Index, developed as part of the BTS Transferability Statistics report and based on census tract location and population density, census tracts were designated as urban, suburban, and rural for purposes of VMT estimation.

Washington and Oregon both developed estimates for daily vehicle miles traveled (DVMT) in support of their on-going road usage charge programs (*3*, *11*). These reports were used to compare estimates developed from the BTS model. The Washington study estimates were within 10% for most counties. The OHAS closely matched the BTS results for urban households but estimated higher DVMT in rural areas than the BTS model. BTS-derived estimates also compared favorably to several California estimates of household DVMT.

These DVMT estimates were for all households in each census tract. Therefore, they included vehicle miles traveled by both gas- and non-gas-powered vehicles. The next step in the study was to estimate the distribution of vehicles by fuel type in each census tract and use this information to estimate aggregate VMT by fuel type.

Table 3. Registration Record Disposition

State	Registration records received	Removed from analysis because: Non-standard passenger vehicle	No fuel type ID	No fuel economy ID	Location rebalance	Final vehicle count by state	Usable vehicle records
Arizona	5,917,640	8%	1%	10%	2%	4,618,996	78.1%
California	27,559,122	17%	3%	0%	1%	21,588,525	78.3%
Idaho	2,746,499	13%	0%	3%	4%	2,194,713	79.9%
Montana	700,000	10%	1%	8%	5%	528,872	75.6%
Oregon	3,782,748	0%	0%	33%	0%	2,524,951	66.7%
Texas	24,203,117	15%	0%	6%	4%	18,047,380	74.6%
Utah	2,330,852	6%	1%	7%	1%	1,979,521	84.9%
Washington	5,130,387	1%	0%	15%	0%	4,315,254	84.1%
Total	72,370,365	13%	1%	6%	2%	55,798,212	77.1%

Table 4. Vehicles by Fuel Type

State	Gas vehicles	Electric/ Hydrogen	Hybrid	Flex- fuel/ Biofuel	Other fossil	Diesel	Total
Arizona	4,135,600	3,233	84,880	339,843	675	54,765	4,618,996
(percentage of total)	89.5%	0.1%	1.8%	7.4%	0.0%	1.2%	100.0%
California	20,563,578	83,213	433,851	267,871	16,508	223,504	21,588,525
(percentage of total)	95.3%	0.4%	2.0%	1.2%	0.1%	1.0%	100.0%
Idaho	2,020,931	290	18,014	106,149	197	49,132	2,194,713
(percentage of total)	92.1%	0.0%	0.8%	4.8%	0.0%	2.2%	100.0%
Montana	420,969	2	6,378	28,813	83	72,627	528,872
(percentage of total)	79.6%	0.0%	1.2%	5.5%	0.0%	13.7%	100.0%
Oregon	2,210,772	14	62,342	112,388	160	139,275	2,524,951
(percentage of total)	87.6%	0.0%	2.5%	4.5%	0.0%	5.5%	100.0%
Texas	14,594,896	27,018	38,190	2,370,422	455,099	561,590	18,047,215
(percentage of total)	80.9%	0.2%	0.2%	13.1%	2.5%	3.1%	100.0%
Utah	1,683,707	1,506	32,797	144,812	2,758	113,941	1,979,521
(percentage of total)	85.1%	0.1%	1.7%	7.3%	0.1%	5.8%	100.0%
Washington	3,983,716	5,084	94,684	140,127	167	91,476	4,315,254
(percentage of total)	92.3%	0.1%	2.2%	3.3%	0.0%	2.1%	100.0%
Eight state total	49,614,169	120,360	771,136	3,510,425	475,647	1,306,310	55,798,047
(percentage of total)	88.9%	0.2%	1.4%	6.3%	0.9%	2.3%	100.0%

Processing and Identifying Fuel Type from Vehicle Registration Information

As shown in Table 3, over 72 million vehicle registration records were received and processed. These records contained VINs and sufficient information to assign the record to a census tract. There were several reasons why vehicle records were removed from the analysis. The primary reason was that registration databases typically contain information for motorcycles, heavy trucks, trailers, motor homes, and other vehicle types excluded from this analysis. Additionally, under the heading of Location Rebalance, vehicles registered outside the targeted states, in census tracts showing no population, or in unrecognizable locations were eliminated.

VINs for the remaining registration information were decoded using the National Highway Safety Administration's (NHTSA) Product Information Catalog and Vehicle Listing (vPIC) database and the Application Programming Interface (API) provided by NHTSA (*12*). The results from vPIC API decoding were used to match EPA fuel economy data to each VIN record (*13*). Losses caused by inability to match fuel type and fuel economy are noted in Table 3.

Once the VINs were decoded, estimating the distribution of fuel types by state and census tract was relatively straightforward. Table 4 shows these distributions for each of the eight states and the percentage of all decoded VINs that these estimates represent. There are important differences between states that can be observed from these data.

Whereas the total number of diesel-fueled vehicles is largest in Texas, the states of Utah, Oregon, and Montana have a higher percentage of registrations when compared with Texas. California has the highest total

Table 5. Estimated Annual VMT Subject to the Gas Tax

State	Urban	Mixed	Rural	Total	Percentage of Total VMT
Arizona	21,102	3,074	1,349	25,525	96.5%
(percentage of total statewide Gas VMT)	82.7%	12.0%	5.3%	100.0%	
California	141,583	7,683	3,203	152,469	99.0%
(percentage of total statewide Gas VMT)	92.9%	5.0%	2.1%	100.0%	
Idaho	4,970	1,234	1,218	7,422	97.4%
(percentage of total statewide Gas VMT)	67.0%	16.6%	16.4%	100.0%	
Montana	2,017	719	1,583	4,319	96.9%
(percentage of total statewide Gas VMT)	46.7%	16.6%	36.7%	100.0%	
Oregon	12,515	2,280	1,151	15,946	97.6%
(percentage of total statewide Gas VMT)	78.5%	14.3%	7.2%	100.0%	
Texas	93,168	15,133	6,059	114,360	92.9%
(percentage of total statewide Gas VMT)	81.5%	13.2%	5.3%	100.0%	
Utah	10,178	492	756	11,426	96.1%
(percentage of total statewide Gas VMT)	89.1%	4.3%	6.6%	100.0%	
Washington	25,164	4,105	1,709	30,978	98.2%
(percentage of total statewide Gas VMT)	81.2%	13.3%	5.5%	100.0%	
Total	310,702	34,721	17,029	362,452	96.5%
(percentage of total statewide Gas VMT)	85.7%	9.6%	4.7%	100.0%	

number of hybrid vehicle registrations, though Oregon and Washington have a higher percentage of hybrid vehicle registrations. Diesel vehicles are not included in this research and hybrids are treated as gas-powered vehicles but tend to have higher average efficiency.

Electric vehicles are primarily registered in California. This state has over 69% of registered electric vehicles in the eight states studied. Most of the flex/biofuel and "other fossil"–fueled vehicles are registered in Texas. Flex-fuel vehicle registrations in Texas are particularly high. These distributions of non-gas-powered vehicles influence the effect of a mileage-based fee because they currently often pay less in gas taxes (or in some cases, no gas taxes).

Estimated VMT for gasoline, hybrid and flex/biofuel vehicles is shown in Table 5 aggregated for the urban, mixed, and rural census tracts in each state, as is the percentage of VMT attributed to gas-powered vehicles in each state. A similar estimate for non-gas-powered vehicles is shown in Table 6. For both tables, some assumptions were necessary. Although gas-powered vehicles can use up to 15% ethanol, and most gasoline sold in the U.S. has up to 10% ethanol, most states tax gasoline with these levels of ethanol at the same rate as pure gasoline. Consequently, the gas-powered vehicle category includes vehicles fueled with up to 10% ethanol.

For flex/biofueled vehicles, motorists have a choice of using either standard-formulated gasoline or an alternative fuel, such as biofuels purchased from non-retail distribution or otherwise untaxed or differentially-taxed fuels. As the choice, especially in rural areas, may be highly dependent on local availability, the assumption made for this study was that 50% of the fuel used by flex/biofueled vehicles was gasoline and 50% was an alternative non-taxed fuel. This assumption is reflected in Tables 5 and 6. It is noteworthy that in many of the more rural states such as Montana and Idaho, there are relatively high concentrations of VMT in the urban areas for non-gas-powered vehicles.

Estimated Effects of Mileage-Based Fees on Rural Households

Effects of a change from a per-gallon tax to a mileage-based fee depend on two key factors: the number of miles driven, and fuel efficiency of the vehicles involved. In this regard, the differences between households located in rural, urban, and mixed census tracts are important to understand, as they directly affect how much households are likely to pay under a revenue-neutral mileage-based fee.

As shown in Table 7, fuel efficiency is consistently higher in urban census tracts compared with rural census tracts. Vehicles registered in mixed census tracts are generally less fuel-efficient than those in UAs, but, like vehicles registered in urban census tracts, are typically more fuel-efficient on average than those in rural areas. Based on registration data, vehicles in rural areas also tend to be older than those in urban areas (by an average of 1.3 years).

The estimates of DVMT by census tract also provided important information on average distances traveled by households in urban, mixed, and rural census tracts. Table 8 shows estimates for DVMT by state. Although rural households travel between 38.6 and 52.6 mi per day, these totals are not a lot greater than urban

Table 6. Estimated Additional VMT Subject to Mileage-Based Fee

State	Urban	Mixed	Rural	Total	Percentage of Total VMT
Arizona	736	118	64	918	3.5%
(percentage of total statewide Non-Gas VMT)	80.2%	12.9%	7.0%	100.0%	
California	1,459	81	33	1,573	1.0%
(percentage of total statewide Non-Gas VMT)	92.8%	5.1%	2.1%	100.0%	
Idaho	127	36	36	199	2.6%
(percentage of total statewide Non-Gas VMT)	63.8%	18.1%	18.1%	100.0%	
Montana	60	20	58	138	3.1%
(percentage of total statewide Non-Gas VMT)	43.5%	14.5%	42.0%	100.0%	
Oregon	287	66	36	389	2.4%
(percentage of total statewide Non-Gas VMT)	73.8%	17.0%	9.3%	100.0%	
Texas	6,502	1,488	726	8,716	7.1%
(percentage of total statewide Non-Gas VMT)	74.6%	17.1%	8.3%	100.0%	
Utah	403	24	42	469	3.9%
(percentage of total statewide Non-Gas VMT)	85.9%	5.1%	9.0%	100.0%	
Washington	433	84	37	554	1.8%
(percentage of total statewide Non-Gas VMT)	78.2%	15.2%	6.7%	100.0%	
Total	10,012	1,918	1,033	12,963	3.5%
(percentage total statewide Non-Gas VMT)	77.2%	14.8%	8.0%	100.0%	

Table 7. Average Fuel Efficiency for Vehicles Paying Gas Taxes (mpg)

State	Urban	Mixed	Rural
Arizona	22.7	22.1	20.9
California	27.0	26.3	25.2
Idaho	21.7	21.2	20.8
Montana	23.8	23.6	22.9
Oregon	21.3	20.3	19.9
Texas	21.6	20.5	19.9
Utah	22.8	21.8	21.1
Washington	22.6	21.5	21.2

Table 8. Estimated Daily VMT per Household in Urban, Mixed, and Rural Census Tracts

	Daily VMT per Household		
State	Urban	Mixed	Rural
Arizona	37.3	45.8	41.4
California	42.1	46.0	39.4
Colorado	41.8	57.7	46.0
Idaho	44.5	52.6	44.8
Montana	41.0	51.6	42.8
Oregon	38.2	42.6	38.6
Texas	47.4	54.1	44.0
Utah	47.7	59.9	52.6
Washington	41.1	46.3	39.7

households. In some states, such as Texas and Washington, urban households travel more on average than rural households. In all states, households in mixed census tracts travel longer distances on average than those in either urban or rural census tracts.

Total gas tax revenues from households was computed using the annualized VMT estimates for gas-powered vehicles at the census tract level, estimated average fuel efficiency based on the composition of vehicles in each census tract, and the state gas taxes per gallon as of 2016. The estimated revenues based on this computation were then used as the base levels to estimate the revenue-neutral mileage-based fee for all drivers that would result in no net gain or loss in revenues. Because each state's per-gallon gas tax is different, the results of these computations are different for each state. Table 9 shows the estimated mileage-based fees that would need to be charged by each state to produce the same amount of revenue as the current gas tax.

The average cost per household in each census tract was then computed using the mileage-based fees shown in Table 9 and census tract VMT estimates. Results were tabulated by the urban, mixed, or rural classification of each census tract. Table 10 shows the reduction in average costs per census tract for each of the states studied. As these are reductions, negative values represent increases in the cost of a mileage-based fee per household compared with the current gas tax and positive values represent savings.

Based on these estimates, households in rural areas would save, on average, between 2% and 6% of what they currently pay in gasoline taxes. Urban households would pay between 0.3% and 1.4% more than they currently pay in gasoline taxes.

These effects are not uniform for all census tracts. There is a wide variation among the census tracts in the

Table 9. Estimated Revenue-Neutral Mileage-Based Fees based on State-Level 2016 Gas Excise Tax Rates

State	Fuel tax	Mileage-based fee
	$ per gallon	$ per mile
Arizona	0.180	0.0077
California	0.300	0.0110
Idaho	0.320	0.0145
Montana	0.270	0.0112
Oregon	0.300	0.0139
Texas	0.200	0.0087
Utah	0.294	0.0125
Washington	0.445	0.0195

Table 10. Reduction in Average Costs under a Mileage-Based Fee for Households in Urban, Mixed, and Rural Census Tracts by State

State	Urban	Mixed	Rural
Arizona	–0.7%	1.7%	6.1%
California	–0.3%	2.4%	6.3%
Idaho	–1.0%	0.9%	3.1%
Montana	–1.4%	–0.4%	1.9%
Oregon	–1.0%	2.9%	4.8%
Texas	–0.5%	1.6%	3.1%
Utah	–0.6%	3.4%	5.5%
Washington	–1.0%	3.6%	4.8%

number of miles driven, the fuel efficiency of the current mix of vehicles in each census tract, and the types of vehicles driven. Figure 3 shows the distribution of costs for a mileage-based fee at the census tract level.

As Figure 3 shows, there are likely to be some rural census tracts where mileage-based fees will increase how much the average household will pay. Likewise, there appear to be many urban census tracts where the average payment per household under a mileage-based fee will decline compared with the current gas tax. What is evident is that neither the additional costs nor the potential savings will be large relative to the current levels of gasoline taxes paid by the average household.

Conclusion

One of the chief concerns expressed about instituting a mileage-based fee is that rural households would be disproportionately affected by a switch to this type of fee structure. Based on this concern, a systematic assessment across multiple western states with varying degrees of urbanization and rural settlement was conducted. The results consistently show that, on average, census tracts that can be characterized as rural under the USDA's RUCA codes are more likely to have their driving costs reduced than increased under a mileage-based fee. The converse is true for UAs. This conclusion is based on data for western states, but the approach used in this study could be applied to states in other regions of the U.S. To the extent that household VMT, fuel type profiles, and household characteristics are similar, these results would be expected to mirror those for the Western U.S.

In the absence of individual household travel information, census tract level household data were used in combination with parameters from the BTS NHTS Transferability Statistics. Each of these sources is subject to statistical errors and the generalizations that are characteristic of national surveys. However, the methods and approach used in this study appear to provide consistent results across states and provide readily explainable and reasonable results. The study also demonstrates the importance of maintaining adequate, accessible vehicle registration data, and the need for developing protocols for accessing this information on an ongoing basis for support of future studies focused on assessing the potential financial impacts of mileage-based fees.

This study required the cooperation of motor vehicle registry departments in each state, which may not always be straightforward to arrange because of concerns about personally identifiable information. However, experience with multiple states in this study illustrates that sufficiently anonymized information can be released so that the basic vehicle identification and geolocation data

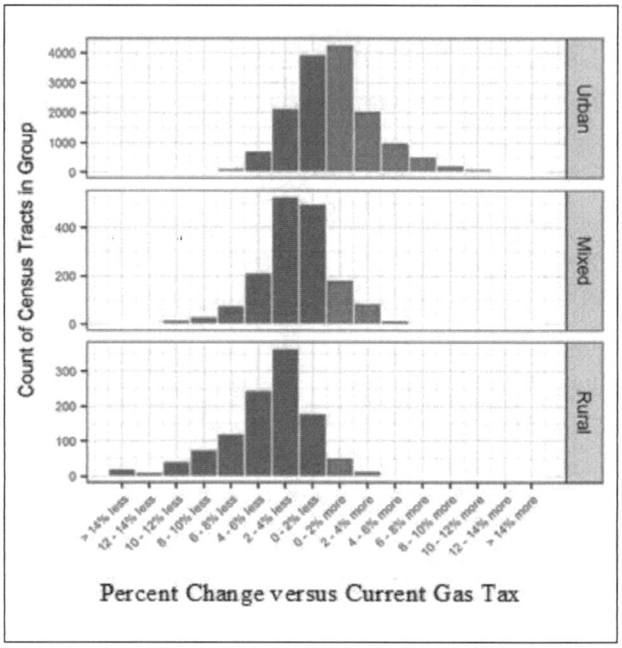

Figure 3. Distribution of effects of a mileage-based fee on households in urban, mixed, and rural census tracts.

needed to resolve information at a census tract level can be obtained.

Acknowledgments

The authors gratefully acknowledge the Western Road Usage Charge Consortium (RUC West) for permission to use data and analysis developed under contract to them. The authors also thank Derek Cutler of EDR Group for database processing support for this paper.

Author Contributions

The authors confirm contribution to the paper as follows: study conception and design: Fitzroy, Schroeckenthaler; data collection: Schroeckenthaler; analysis and interpretation of results: Fitzroy, Schroeckenthaler; draft manuscript preparation: Fitzroy. All authors reviewed the results and approved the final version of the manuscript.

References

1. Kirk, R. S., and W. J. Mallett. *Funding and Financing Highways and Public Transportation.* Congressional Research Service Report R44674. Congressional Research Service, Washington, D.C., 2016.
2. Transportation Research Board. *TRB Special Report 303: Equity of Evolving Transportation Finance Mechanisms.* The National Academies Press, Washington, D.C., 2011. https://doi.org/10.17226/13240.
3. McMullen, B. S., and H. Wang. *Road Usage Charge Economic Analysis.* Final Report SPR 774. Oregon Department of Transportation, 2016. http://www.oregon.gov/ODOT/Programs/ResearchDocuments/SPR774_RoadUsageCharge_Final.pdf.
4. Larsen, L., M. Burris, D. Pearson, and P. Ellis. Equity Evaluation of Fees for Vehicle Miles Traveled in Texas, *Transportation Research Record: Journal of the Transportation Research Board*, 2012. 2297: 11–20.
5. Zhang, L., B. S. McMullen, D. Valluri, and K. Nakahara. Vehicle Mileage Fee on Income and Spatial Equity. *Transportation Research Record: Journal of the Transportation Research Board*, 2009. 2115: 110–118.
6. Paz, A., A. Nordland, and N. Veeramisti. Assessment of Economic Impacts of a Vehicle Mile Traveled Fee for Passenger Vehicles in Nevada. *Transportation Research Record: Journal of the Transportation Research Board*, 2014. 2450: 26–35.
7. Kastrouni, E., K. Gkritza, S. Hallmark, and R. Stephanson. Fuel Tax Versus Vehicle-Miles-Traveled Fee: Identifying Vulnerable Households by Three-Stage Least Squares Analysis. *Transportation Research Record: Journal of the Transportation Research Board*, 2015. 2531: 161–169
8. Weatherford, B. Distributional Implications of Replacing the Federal Fuel Tax with Per Mile User Charges. *Transportation Research Record: Journal of the Transportation Research Board*, 2011. 2221: 19–26.
9. Crandall, M., and B. Weber. *Defining Rural Oregon: An Exploration.* Paper RSP-05-03. Oregon State University, Rural Studies Program. https://secure.engr.oregonstate.edu/wiki/transportation/uploads/ODOT07-VMT/defining%20rural.pdf. Accessed November 9, 2017.
10. Bureau of Transportation Statistics. Methodology – Local Area Transportation Characteristics for Households. https://www.bts.gov/archive/subject_areas/national_household_travel_survey/methodology. Accessed July 23, 2018.
11. Washington State Transportation Commission. *Road Usage Charge Assessment: Financial and Equity Implications for Urban and Rural Drivers.* Washington State Transportation Commission, 2015. https://waroadusagecharge.files.wordpress.com/2014/05/2015-0227_urbanruralreport.pdf.
12. *National Highway Traffic Safety Administration. VIN Decoder. Product Information Catalog and Vehicle Listing (vPIC).* https://vpic.nhtsa.dot.gov/. Accessed July 26, 2017.
13. *EPA Fuel Economy Database.* https://www.fueleconomy.gov/feg/download.shtml. Accessed July 26, 2017.

The Standing Committee on Revenue and Finance (ABE10) peer-reviewed this paper (18-03534).

Opinions expressed in this paper are the views of the authors and are not representative of RUC West or any of their member states.

Article

TRR
JOURNAL OF THE TRANSPORTATION RESEARCH BOARD

Transportation Research Record
2018, Vol. 2672(4) 32–42
© National Academy of Sciences:
Transportation Research Board 2018
Article reuse guidelines:
sagepub.com/journals-permissions
DOI: 10.1177/0361198118796737
journals.sagepub.com/home/trr
SAGE

Implication of Replacing the Federal and State Fuel Taxes with a National Vehicle Miles Traveled Tax

Yiwei Wang[1] and Qing Miao[2]

Abstract

A vehicle miles traveled (VMT) tax is often proposed to replace fuel taxes for financing the nation's highway and road network. In this paper, we investigate households' driving response to driving cost changes depending on their vehicle choices. Using the empirical estimates, we simulate the vehicle usage, tax burdens, and total tax revenues generated under a possible nationwide revenue-neutral flat VMT tax. Our results confirm that, compared with the current gasoline tax, a revenue-neutral flat VMT tax can be a more stable tax revenue source. We estimate that a 50% increase in average miles per gallon would lead to a 28% decrease in the total revenues raised by the current gasoline tax, while the same amount of increase in fuel economy would increase the VMT tax revenues by 4.4 % (all relative to the 2009 baseline). In the meantime, we find no significant difference between the two types of tax in their total revenues, when the pre-tax gasoline prices fluctuate by different magnitudes. A VMT tax would be slightly more regressive than the gasoline tax, but the difference is negligible. Overall, our simulation shows that VMT tax could serve as a viable alternative to gasoline taxes.

In the past few decades, the U.S. federal and state governments have been facing growing pressure to finance their transportation infrastructures with existing fuel taxes (*1*, *2*). Levied on the cents-per-gallon basis, fuel taxes generate insufficient revenue for funding the nation's highway and road network, when fuel consumption persistently declines due to the improved vehicle fuel economy (*2*). According to the U.S. Department of Transportation (DOT), the Highway Trust Fund (HTF) that is primarily financed by the federal fuel tax revenue is nearing insolvency, and a recent HTF shortage in 2014 was averted by budget transfers from the general account. Many predict that the financing gap will keep widening in the coming decades as the renewed Corporate Average Fuel Economy (CAFE) standards require even higher levels of fuel economy.

One option to remedy the highway financing problem is to replace the gasoline tax with a vehicle miles traveled (VMT) tax, which charges people based on the distance they drive instead of the amount of gasoline they consume (*3*, *4*). Because the aggregate VMT normally grows faster than fuel consumption, the VMT tax is expected to provide a more stable revenue stream (*1*). Such a tax would also arguably improve the benefits equity as it makes all drivers pay the same rate per mile for traveling and those who use the public road system more pay more fees (*5*, *6*).

Oregon is the first state in the United States that adopted a voluntary VMT tax program in 2014 after implementing several pilot projects. While Oregon's new policy initiative has renewed the debate on the feasibility of VMT tax, less research has examined the potential behavioral responses to this tax policy as well as its fiscal and distributional implications. In this paper, we estimate how a flat VMT tax would alter households' driving behaviors. We further project its revenue-generating potential, allowing for continued fuel economy improvement, and the tax burden borne by different income groups.

Our research proceeds in two steps. In the first phase, we use the 2009 National Household Transportation Survey (NHTS) data to estimate the price sensitivity of driving demand, depending on a household's vehicle choices. We find that households with multiple vehicles are more sensitive to driving cost changes compared with those with only one vehicle. Households that own SUVs

[1]Center for Governmental Research, Rochester, NY
[2]Department of Public Policy, Rochester Institute of Technology, Rochester, NY

Corresponding Author:
Address correspondence to Qing Miao: qxmgla@rit.edu

and vans are more sensitive to driving cost fluctuations than those with smaller passenger cars.

We then use our empirical estimates to quantify the changes in annual VMT, tax revenues, and tax burden, assuming that the gasoline tax is replaced with a nationwide flat VMT tax. We solve the VMT tax rate for each state that would raise the equivalent amount of revenue as their gasoline taxes do. Our simulation results confirm that a VMT tax would generate much more stable revenues than a gasoline tax does, especially when the average fuel economy increases significantly. Pre-tax gasoline price fluctuation would cause similar changes in both gasoline and VMT tax revenues. A VMT tax would be slightly more regressive than the gasoline tax, but the difference is negligible.

This research contributes to a nascent literature on the VMT tax (*6–9*) by estimating its fiscal and distributional implications at the national scale. By incorporating the future improvement of fuel economy into the analysis, we provide useful parameters for policymakers to estimate the revenue-generating potential of a VMT tax. Our research also contributes to the VMT literature by examining the heterogeneity of driving demand conditional on households' vehicle choices (*10*, *11*).

Policy Background

Much of the discussion on VMT tax focuses on the comparison between the current fuel tax and VMT tax from different perspectives. Many economists believe that raising fuel taxes to address the transportation financing gap is the optimal solution because a higher tax rate would not only increase tax revenues, but also internalize the externality of driving which includes the social marginal costs of emissions, traffic congestion, and accidents. Parry and Small (2005) estimate that the optimal U.S. gasoline tax should be set at $1.01 per gallon so that it could internalize all the externalities of driving (*12*). However, some argue that fuel taxes do not efficiently capture the damage caused by vehicle travel to the road system, considering the differences in vehicle weight and fuel economy (*7*).

The new CAFE standards introduced by the Obama administration require the average fuel economy of new passenger cars to almost double from 2010 to 2025, and that of light trucks to increase by almost 70% from its 2010 level. With the projected increase in average fleet fuel economy, the federal and state governments might have to raise the fuel tax rates substantially to compensate their potential revenue losses. Whether this solution is politically feasible, however, is debatable. The federal gasoline tax has been fixed at 18.4 cents since 1993, regardless of the inflation in expenditures, which implies a more than 30% reduction in real purchasing power (*1*).

The VMT tax is one type of road pricing, set at a per-mile-based rate which normally does not vary by vehicle type. Dozens of pilot programs of mileage-based charges were implemented across the states in the late 2000s and early 2010s to assess the feasibility of this new tax mechanism. Studies on these pilot programs suggest that many participants had a more positive view of the mileage-based tax after experiencing them (*2*, *13*). However, introducing a VMT tax still faces public acceptance challenges, including the privacy issue and higher administration cost of installing the tracking device (*14*, *15*).

In 2013, Oregon passed Senator Bill 810 which established the nation's first mileage-based revenue program for light-duty vehicles. A GPS tracking device was installed on volunteered vehicles (limited to 5,000 cars and light-duty commercial vehicles in the first phase) at the cost of approximately $250 per vehicle. Drivers were charged at a rate of 1.5 cents per mile of their traveled distance on public roads, regardless of their vehicle type and model. Participants of the program received monthly bills of their road-use charges and had the state gasoline tax refunded when they purchased gasoline at pumps in Oregon. In July 2016, California also started implementing a similar road charging pilot project.

Since the VMT tax is a relatively new concept and still in the experimental stage, less is known about its actual impact on driving behaviors, tax revenues, and tax burden. Several studies have examined its distributional implications. McMullen et al. (2010) find that a VMT tax (set at 1.2 cents per mile) that replaces a 24-cents-per-gallon gasoline tax would be slightly more regressive than the fuel tax (*7*). The VMT tax would benefit the rural households more than their urban counterparts because the vehicles owned by rural households are on average less fuel-efficient. Zhang et al. (2009) estimate that Oregon's flat VMT tax would not lead to significant changes in the tax burden and taxpayers' welfare in either the short or long term (*16*). Weatherford (2011) finds that a VMT tax, set at 0.98 cents per mile, would shift the tax burden from the low-income to high-income households, from rural to urban households, and from retired to younger households with children (*6*). Robitaille et al. (2011) compare a $ 0.015 VMT tax with a $ 0.10 federal gasoline tax increase, and conclude that the former leads to a larger decrease in consumer surplus and loss of social welfare (*8*).

Regarding the environmental impact, Greene (2011) argues that a VMT tax does not encourage improvement in fuel economy and thus does not provide additional environmental benefits (*9*). Caplan (2009) estimates that a VMT tax (set at 0.3 cents per mile for cars and 1 cent per mile for light trucks) would reduce the annual particulate emissions by 7%–11%, suggesting a positive environmental effect (*17*).

It is important to note that when simulating the impacts of a VMT tax, one needs to consider how driving demand would respond to changes in the tax structure and driving costs. Our paper also draws upon the literature on modeling driving behaviors, including studies estimating the price elasticity of VMT and how VMT reacts to changing fuel efficiency. As for the latter, economic theories suggest that improved vehicle fuel economy reduces the fuel cost of travel per mile, and may cause the "rebound effect" (i.e., increased driving demand offsets some of the energy-saving benefits) (*18, 19*). The magnitude of the rebound effect captures the sensitivity of vehicle travel to improved fuel economy and thus should also be factored into the calculation of VMT tax and fuel tax revenues.

Data

We compile the data for this study from various sources, which we describe as below.

National Household Travel Survey

To empirically model driving demand, we use the 2009 NHTS, which is the most recent nationwide travel survey conducted by the Federal Highway Administration (FHWA) of DOT. The 2009 NHTS sample includes 25,510 households in the original national sample and 124,637 households in the add-on sample. The sampled households were asked to provide detailed information about their owned vehicles (e.g., vehicle type, make, model and model year), their annual usage (or VMT), costs of fuel purchase, fuel prices they faced, and types of fuels. The data also contain socioeconomic and demographic information of the sampled household members (e.g., income, educational background, and occupation).

Vehicle Attributes Data

We combine the Ward's Automotive Yearbooks (WAY) and Automobile Catalog (AC) to construct a data set of vehicle attributes, which include vehicle fuel economy (MPG) and vehicle class. We merge the NHTS data with our compiled vehicle attributes based on individual vehicle model and model year. It should be noted that the NHTS data contain a variable indicating MPG appended by the Energy Information Administration (EIA). However, we prefer to use the Environmental Protection Agency (EPA) fuel efficiency rating (the measure used in our compiled attribute data) because this is the information disclosed to consumers when they purchase vehicles and is thus closer to the perceived fuel efficiency level that would directly influence households' driving behaviors.

In the final data set, we match more than 90% of the passenger vehicles in the 2009 NHTS data with our vehicle attribute data. Our final sample includes a total of 94,645 households and 207,382 vehicles in 48 states. We do not include Hawaii and Iowa because of their incomplete regional and tax rate information.

As shown in Table 1, in households with multiple vehicles, the fleet consisted of a relatively high percentage of light-duty trucks, including pickups, sport-utility vehicles (SUVs) and vans, whereas more than half of the vehicles in single-vehicle households are from the small cars category. The average fuel economy is slightly higher in single-vehicle households than in all vehicle categories.

Vehicle choices also correlate with household incomes. As shown in Table 2, over 70% of the households with annual incomes below $10,000 have only one vehicle. By contrast, over 94% of the households with annual incomes above $100,000 own multiple vehicles. Regarding vehicle type, lower-income households are also more likely to own small cars than high-income households.

Model

To project the potential driving response to a revenue-neutral VMT tax, we first estimate the sensitivity of driving demand (measured by annual household VMT) to fuel cost changes. We first specify the following model using our cross-sectional data with *i* indexing individual vehicle and *j* indexing household as:

$$M_{ij} = \beta\ GP_j + GP_j{*}T_{ij}'B_1 + T_i'B_2 + X_j'B_3 + e_{ij}, \quad (1)$$

Table 1. Vehicle Composition in Single-Vehicle and Multiple-Vehicle Households

	Single-vehicle households			Multiple-vehicle households		
Vehicle type	Count	Percentage	Mean MPG	Count	Percentage	Mean MPG
Small cars	17,105	54.9%	27.4	69,005	39.2%	27.1
Large cars	4,273	13.7%	21.9	14,802	8.4%	21.8
Pickups	2,285	7.3%	20.0	38,294	21.7%	19.5
SUVs	5,217	16.7%	20.9	39,381	22.3%	19.8
Vans	2,294	7.4%	20.5	14,726	8.4%	20.3
Total	31,174		24.5	176,208		22.8

Table 2. Vehicle Distribution across Income Groups

Household annual income ($)	Number of households	% of single-vehicle households	Percentage of vehicle type				
			Small cars	Large cars	Pickups	SUVs	Vans
<5,000	1025	70.2%	47.9%	12.5%	17.6%	14.0%	8.1%
5,000–9,999	2437	72.0%	49.8%	12.4%	18.3%	11.5%	8.0%
10,000–14,999	4158	69.4%	47.9%	12.5%	18.8%	13.3%	7.5%
15,000–19,999	4826	61.5%	45.0%	13.2%	19.7%	13.8%	8.4%
20,000–24,999	4349	56.9%	45.0%	12.7%	19.5%	14.6%	8.2%
25,000–29,999	6385	46.8%	41.3%	12.9%	21.6%	15.6%	8.6%
30,000–34,999	3774	42.7%	41.0%	11.7%	21.8%	16.6%	8.9%
35,000–39,999	6216	36.1%	39.9%	11.5%	22.4%	17.6%	8.6%
40,000–44,999	3242	33.0%	39.8%	10.0%	22.3%	19.0%	8.9%
45,000–49,999	6222	28.6%	39.0%	10.9%	23.1%	18.8%	8.2%
50,000–54,999	2924	26.5%	40.0%	9.2%	22.6%	20.0%	8.2%
55,000–59,999	5623	21.5%	39.6%	9.3%	22.3%	20.7%	8.2%
60,000–64,999	2198	19.2%	40.0%	8.0%	22.3%	21.3%	8.4%
65,000–69,999	4843	16.9%	39.2%	9.1%	21.0%	22.3%	8.4%
70,000–74,999	2250	14.7%	39.5%	7.7%	21.7%	23.0%	8.2%
75,000–79,999	4592	13.2%	40.1%	8.0%	20.7%	23.2%	8.0%
80,000–99,999	9231	9.2%	39.8%	6.7%	19.9%	24.5%	9.1%
≥100,000	20350	5.9%	42.9%	6.6%	14.9%	28.1%	7.5%

where

M_{ij} = annual VMT on individual vehicle i in household j

GP_j = fuel cost a household faces (self-reported average after-tax fuel cost of households, measured by dollars per gallon)

T_{ij} = a set of dummy variables indicating the vehicle type (e.g., small cars, large cars, pickups, SUVs, and vans)

We interact T_{ij} with GP_j to examine the heterogeneity in driving demand conditional on vehicle types. X_j denotes a household's characteristics, including its annual income, educational background, race, and age of the household respondents, number of workers, number of drivers, reported distance to work, as well as its locational information including the population density of its residence areas, the size of the metropolitan statistical area (MSA), whether the household is located in a rural area, and whether there is rail service in that area. We cluster standard errors at the household level to address possible heteroskedasticity.

Results

Empirical Estimates: Driving Responses to Fuel Cost Changes

Table 3 reports the regression results for households with one vehicle and households with multiple vehicles, respectively. In Columns 1 and 3 where we do not include the interaction terms, the gasoline price coefficients should be interpreted as the average change in annual VMT when a household faces a $1 increase in per gallon gasoline cost. Comparing their magnitudes, we find that single-vehicle households are less sensitive to fuel cost changes than households with multiple vehicles. This result is not surprising because the latter has more traveling options; for example, when driving becomes more expensive couples may choose to drive one car to go to work instead of driving separate vehicles. These households may deliberately use the more efficient vehicle more often when the gasoline price increases.

In Columns 2 and 4 where we include the interaction terms, the gas price coefficients represent the average effect of $1 gas price increase on VMT using small cars (the omitted category). The VMT price sensitivity for other vehicle types is the sum of the baseline gas price coefficient and their respective interaction terms. Table 4 reports our computed estimates for each vehicle type. We find that multiple-vehicle households respond significantly to gasoline cost changes in all vehicle categories, with particularly larger responses in large cars and vans. This is probably because large cars, SUVs, and vans are used more often for recreational purposes, which are more flexible in travel demand, whereas small passenger cars are used more often for work-related commuting, which is more price inelastic. In comparison, single-vehicle households only reduce their VMT on SUVs significantly when the fuel cost increases. Overall, our results suggest that households' driving responses to gasoline costs vary by the number and the type of vehicles

Table 3. Household Annual VMT Response to Fuel Cost Changes

	Single vehicle		Multiple vehicle	
Variable	(1)	(2)	(3)	(4)
Gas price	−945.0***	−666.0	−1378.0***	−636.4***
	(304.7)	(375.3)	(158.1)	(227.9)
Large car		−1637.6		3357.8**
		(2627.5)		(1702.9)
Pickup		1552.4		757.9
		(3409.8)		(1077.8)
SUV		7021.1***		3262.6***
		(2465.1)		(1081.9)
Van		461.4		9353.4***
		(3598.6)		(1595.6)
GP*Large car		515.1		−1024.0
		(858.2)		(557.9)
GP*Pickup		−174.1		−478.1
		(1108.4)		(350.6)
GP*SUV		−2049.3**		−806.4**
		(797.4)		(351.7)
GP*Van		40.2		−2754.2***
		(1169.0)		(518.5)
Log population density	−515.4***	−497.6***	−372.3***	−388.2***
	(35.4)	(35.3)	(19.5)	(19.6)
Number of workers	197.4	186.7	−107.7***	−107.8***
	(128.4)	(128.4)	(36.2)	(36.2)
Household size	526.9***	486.4***	229.2***	176.5***
	(148.2)	(149.3)	(42.8)	(43.1)
Number of drivers	1081.3***	1078.0***	115.5**	176.0***
	(139.1)	(138.9)	(58.3)	(58.5)
Distance to work	108.1***	108.0***	68.1***	69.5***
	(34.0)	(34.2)	(6.3)	(6.4)
Rail in area	−651.9***	−571.3***	−446.4***	−512.7***
	(154.3)	(154.4)	(86.7)	(87.5)
Rural	619.9***	571.7***	382.9***	408.9***
	(128.0)	(128.0)	(62.8)	(62.8)
Male household respondent	1368.4***	1266.7***	192.4***	202.7***
	(88.7)	(90.3)	(41.5)	(41.5)
N	31174	31174	176208	176208
R-squared	0.1867	0.1843	0.1037	0.0993

Note: Values in parentheses are standard errors clustered at the household level. All the specifications include dummy variables that control for household income, age, race, and educational background of household head, life styles, and the size of MSAs where households are located.
$^{*}p < 0.1$. $^{**}p < 0.05$. $^{***}p < 0.01$.

they own. Accounting for the heterogeneous VMT responses is crucial for estimating the effect of a flat VMT tax, which imposes differential driving costs depending on the type of vehicles.

As for other control variables, we find that households living in more densely populated areas or having access to rail transportation drive significantly less. Larger households, those facing longer distances to work or living in rural areas report significantly higher VMT. All these results suggest that households' driving behavior highly correlates with their demand for mobility and the purpose of travel. To examine the robustness of our empirical results, we also used a log-log model to estimate the price elasticity of driving demand. Our results (available upon request) are highly consistent with the level model.

Simulation Analyses

In this section, we first solve for the state-specific VMT tax rates that would generate the same amount of revenues as the federal and state gasoline taxes do. Based on our estimated VMT taxes, we then simulate their effects on gasoline consumption, vehicle usage, and tax revenues under various scenarios, including increased average fleet fuel economy (as required by the renewed CAFE standards), fluctuations in pre-tax gasoline prices, and increased gasoline tax rate.

Replacing the Gasoline Tax with a Revenue-Neutral Flat VMT Tax. As we discussed above, the VMT tax is known as a more stable tax structure and less susceptible to increased fuel

Table 4. Changes in Annual VMT with Respect to $1 Increase in Gasoline Costs

Vehicle type	Single vehicle	Multiple vehicle
Small/medium car	−665.98	−636.42***
	(375.29)	(227.89)
Large car	−150.93	−1660.39***
	(791.05)	(523.06)
Pickup	−840.05	−1114.47***
	(1053.05)	(282.82)
SUV	−2715.31***	−1442.82***
	(726.30)	(302.45)
Van	−625.80	−3390.66***
	(1123.97)	(488.53)

Note: The response rates of small cars are the coefficients of gasoline price in Table 2. The response rates of other vehicles are the linear combinations of gasoline price and the interaction terms of gasoline price and corresponding vehicle category dummies in Table 2.

economy. To quantify its revenue-generating potential, we assume that each state sets its VMT tax rate such that it would generate the same amount of tax revenues as the gasoline tax does. In doing so, we first calculate the total gasoline consumption of our sampled households in state s using their reported annual usage of each vehicle j and the corresponding MPG level (i.e., gasoline consumption$_s$ $= \sum(\mathrm{VMT}_{sij}/\mathrm{MPG}_{sij})$. We then compute the sum of federal and state gasoline tax revenues collected in each state (GTR_s) using the per-gallon gasoline tax rates (sum of state and federal tax rate, measured in dollars per gallon, denoted as $\mathrm{T}_s + \mathrm{T}_f$) in 2009 and the sum of gasoline consumption. We specify our equation as:

$$\mathrm{GTR}_s = (\mathrm{T}_s + \mathrm{T}_f)*\sum(\mathrm{VMT}_{sij}/\mathrm{MPG}_{sij}). \quad (2)$$

In theory, we could solve for the state-specific VMT tax rates by dividing the total gasoline tax revenue in each state (GTR_s) by the annual total VMT. However, two issues arise here. First, the annual VMT should be a function of the VMT tax rate, because the new fuel costs that incorporate VMT taxes would affect household's driving demand. Note that we convert the VMT tax rate into the per-gallon driving cost and add that to the pre-tax gasoline price. Assuming that the average price sensitivity of driving demand remains constant, we apply our empirical estimate derived from the last section to calculate the annual VMT.

Second, unlike the per-gallon gasoline tax, a flat VMT tax would incur different driving costs for various types of vehicles using the same amount of fuel. For example, a household has two vehicles: a car and an SUV. Suppose that the car has an MPG of 30, the SUV has an MPG of 15, and the gasoline price is $3.00 per gallon, which includes 30 cents of gasoline tax (i.e., pre-tax gasoline price is $2.7 per gallon). Now the government decides to replace the gasoline tax with a VMT tax for which the rate is set at 1.5 cents per mile. Under this new tax mechanism, to drive 30 miles using the small car, the household will pay $3.15 (for gasoline plus tax), compared to $3.00 under the gasoline tax. If the household chooses to drive 15 miles using the SUV, they will pay $2.925, whereas the cost would still be $3.00 under the gasoline tax. Therefore, in this case, the tax paid under a VMT tax would increase for using the car and decrease for the SUV, when both vehicles consume one gallon of gasoline. It suggests that the implicit cost of VMT tax imposed on gasoline consumption is higher for more fuel-efficient vehicles because they can travel more miles using the same amount of fuel.

We specify that for vehicle i in household j, the cost of using one gallon of gasoline (or the equivalent per-gallon gasoline cost) under a VMT tax, VG_{ij}, is equal to GP_{ij} - $(\mathrm{T}_s + \mathrm{T}_f) + \mathrm{MPG}_{ij}*\mathrm{VT}_s$. GP_{ij} denotes the current after-tax gasoline price. VT_s is the VMT tax rate adopted in state s, measured in dollars per mile. It should be noted that we assume the pre-tax gasoline prices, as well as the vehicle stocks, remain the same. The annual VMT under the new VMT tax, VMT_{ij}^*, is a linear function of VG_{ij}. We then estimate the new annual VMT of vehicles in the state s by applying Equation 2 and our empirical estimates of the price sensitivity of driving demand. Hence, we derive the equation for the VMT tax revenue in a state, shown as below, and equalize that to its total gasoline tax revenues. Finally, we solve the VMT tax rate (VT_s) in each state using a greedy algorithm:

$$\overbrace{\sum\{\beta[\mathrm{GP}_{ij} - (\mathrm{T}_s + \mathrm{T}_f) + \mathrm{MPG}_{ij}*\mathrm{VT}_s] + [\mathrm{GP}_{ij} - (\mathrm{T}_s + \mathrm{T}_f) + \mathrm{MPG}_{ij}*\mathrm{VT}_s]*\mathrm{T}_i'\mathrm{B}_1 + \mathrm{T}_i'\mathrm{B}_2 + \mathrm{X}_j'\mathrm{B}_3\}}^{\text{Total annual VMT in state } s \text{ under its VMT tax}}*\mathbf{VT}_s = \mathbf{GTR}_s \quad (3)$$

Table 5 lists the VMT tax rates we compute for each state that replaces the federal and state gasoline taxes. Note that we use the federal and states' gasoline tax rates in 2009, which correspond to the year of the NHTS survey data we used in our empirical analysis. The highest VMT tax is 2.344 cents per mile in West Virginia, while the lowest is 1.221 cents per mile in Georgia. Under the new VMT tax, the average annual VMT per vehicle would drop by 0.11% to 10,016 miles, and the average gasoline consumption would increase by 0.71% to 457.95 gallons.

Table 5. State-Specific VMT Tax Rates Replacing the 2009 Gasoline Taxes

State	2009 state gas tax (cent per gallon)	2009 state & federal gas tax (cent per gallon)	VMT tax (cent per mile)
AL	18	36.4	1.617
AR	18	36.4	1.982
AZ	21.5	39.9	1.600
CA	18	36.4	1.609
CO	22	40.4	1.680
CT	25	43.4	1.757
DE	23	41.4	1.848
FL	15.6	34	1.599
GA	7.5	25.9	1.221
IA	25	43.4	1.654
ID	19	37.4	1.780
IL	18	36.4	1.653
IN	21	39.4	1.592
KS	24	42.4	1.870
KY	22.5	40.9	1.985
LA	20	38.4	2.003
MA	28.4	46.8	1.737
MD	23.5	41.9	1.779
ME	21	39.4	2.118
MI	19	37.4	1.783
MN	22.5	40.9	1.941
MO	18.4	36.8	1.501
MS	17	35.4	2.200
MT	27.75	46.15	1.968
NC	26	44.4	2.202
ND	24	42.4	1.842
NE	19.6	38	1.861
NH	10.5	28.9	1.727
NJ	18.875	37.275	1.262
NM	24.45	42.85	1.554
NV	30.15	48.55	1.779
NY	23	41.4	1.886
OH	28	46.4	2.118
OK	17	35.4	1.633
OR	24	42.4	1.566
PA	30	48.4	1.877
RI	30	48.4	2.207
SC	16	34.4	1.628
SD	22	40.4	1.767
TN	20	38.4	1.787
TX	20	38.4	1.898
UT	24.5	42.9	1.772
VA	21	39.4	1.543
VT	17.5	35.9	1.731
WA	37.5	55.9	2.077
WI	32.2	50.6	2.208
WV	30.9	49.3	2.344
WY	14	32.4	1.474

This new tax would affect households differently given their vehicle ownership and choices. As we have shown earlier, single-vehicle households have a relatively static driving demand when facing driving cost changes, whereas multiple-vehicle households are more responsive to cost changes.

In Table 6, we examine the impact of VMT taxes (based on our calculated rates) on the average tax payments across various income groups as compared with the gasoline tax. Our simulation indicates that under a VMT tax, low-income households would bear a slightly heavier tax burden, which implies the flat VMT tax might be more regressive than the gasoline tax. However, we note that the difference between the two taxes is negligible, in terms of both absolute values and percentages. It is also important to note that many states currently use revenues from the general sales tax to fill the funding shortfall caused by the declining gasoline tax revenues, and this approach would worsen the regressivity. Within this context, a VMT tax could serve as a better policy option regarding the distributional implication.

Fiscal Implications of Increased Average Fleet Fuel Economy. Because the renewed CAFE standards require a rapid increase in the average fuel economy of new light-duty vehicles by 2025, it may further reduce the gasoline tax base and funding for transportation infrastructures. It should be noted that under either VMT tax or gasoline tax, increases in the overall fuel economy would decrease per mile driving cost and might trigger the rebound effect. If households drive more as fuel economy improves, they will pay more in VMT tax. Nonetheless, households would pay less gasoline tax, because the increased fuel economy lowers their gasoline consumption, and the rebound effect on increased driving might not be sufficient to offset their overall reduced gasoline consumption.

In Figure 1, we report the estimated changes in total tax revenues of both the gasoline tax and VMT tax, when the average fleet fuel economy in the U.S. increases by 10%–50%. Our simulation is based on the assumption that the pre-tax gasoline price remains stable and households are fully aware of the changes in their actual driving costs due to taxes. We show that annual VMT would increase under both two taxes, as a result of improved fuel economy. In particular, the annual VMT is projected to increase by 6.5% under the current gasoline tax, if the average MPG increased by 50%. Meanwhile, the gasoline consumption and tax revenue would drop by 28.4% (as compared with the 2009 baseline). In the same scenario, a flat VMT tax scheme would increase the annual VMT by 4.4% (as compared with 2009 baseline), and the VMT tax revenue would increase in par with the change in driving demand.

Stability under Pre-Tax Gasoline Price Fluctuations. Next, we relax the assumption that pre-tax gasoline prices remain stable and examine the impact of fluctuating gasoline prices on the annual VMT and revenues of a VMT tax and gasoline tax. In Table 7, we report our

Table 6. Tax Burden for Various Income Groups under the VMT Tax

Household income	Mean annual gasoline tax burden	Mean annual VMT tax burden	Mean change in tax burden	Percentage change
under 5,000	154.02	158.05	4.02	2.61%
5,000–9,999	135.31	140.86	5.55	4.10%
10,000–14,999	143.93	150.83	6.90	4.79%
15,000–19,999	167.15	174.76	7.61	4.55%
20,000–24,999	194.01	199.48	5.47	2.82%
25,000–29,999	221.70	225.96	4.26	1.92%
30,000–34,999	251.14	254.80	3.66	1.46%
35,000–39,999	273.11	275.26	2.15	0.79%
40,000–44,999	296.70	296.27	–0.43	–0.15%
45,000–49,999	310.95	313.00	2.05	0.66%
50,000–54,999	340.11	343.53	3.42	1.01%
55,000–59,999	355.65	356.65	1.00	0.28%
60,000–64,999	379.89	381.37	1.48	0.39%
65,000–69,999	385.13	384.94	–0.19	–0.05%
70,000–74,999	408.94	405.79	–3.15	–0.77%
75,000–79,999	417.01	415.18	–1.84	–0.44%
80,000–99,999	451.42	449.92	–1.50	–0.33%
above 100,000	491.37	484.91	–6.45	–1.31%

Note: All the dollar values are adjusted at the 2009 level.

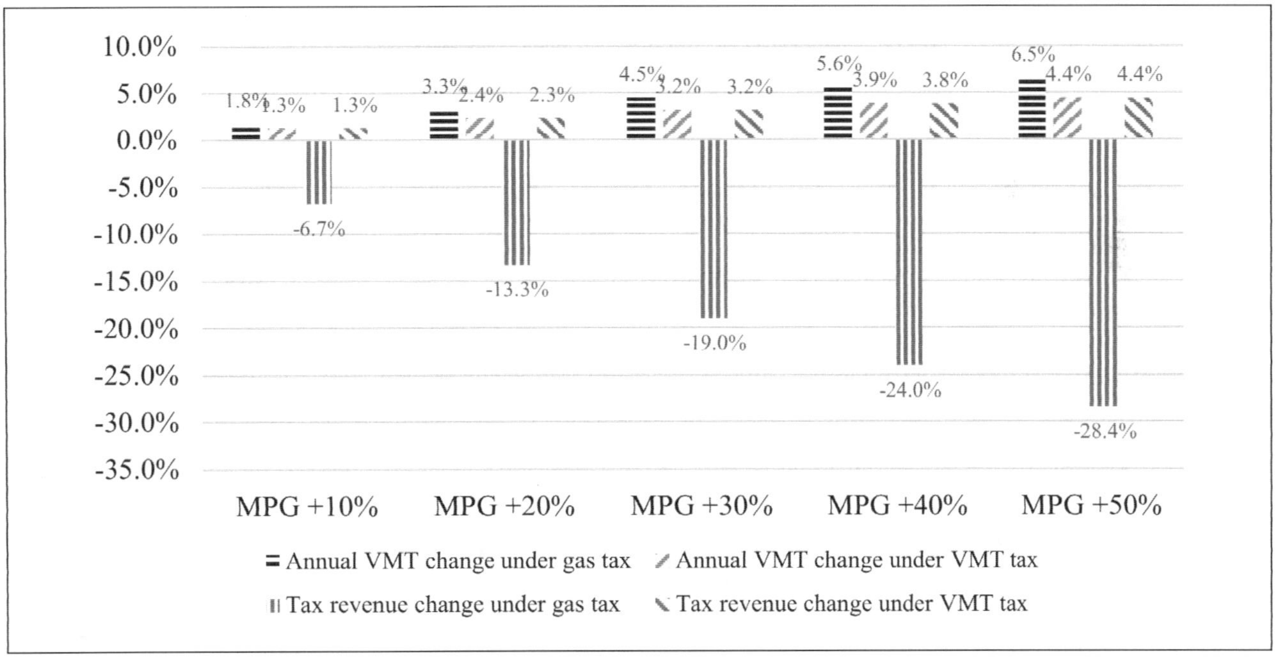

Figure 1. Percentage changes in VMT and tax revenue with respect to fleet MPG changes.
Note: The baseline is the gasoline tax revenue and total vehicle usage in the 2009 NHTS data sample. All percentage changes reflect changes in comparison to the 2009 baseline.

simulation results when the pre-tax gasoline price changes from −50% to + 100%. We show that gasoline price fluctuation does not result in significant differences between the two types of taxes. Revenues generated by a VMT tax would be slightly less than the gasoline tax revenue when the pre-tax gasoline price fluctuates, but the differences would be less than one percent in all cases.

Table 7. Tax Revenues under Pre-Tax Gasoline Price Fluctuation

Pre-tax gas price change	Variable	Gas tax	VMT tax	Change from gasoline tax (%)
0%	Mean annual VMT per vehicle	10028	10016	–0.11%
	Mean gas consumption per vehicle	454.73	457.95	0.71%
	Mean tax payment per vehicle	173.46	173.46	0.00%
−20%	Mean annual VMT per vehicle	10372	10361	–0.11%
	Mean gas consumption per vehicle	473.56	473.60	0.01%
	Mean tax payment per vehicle	180.80	179.39	–0.78%
−50%	Mean annual VMT per vehicle	10888	10877	–0.10%
	Mean gas consumption per vehicle	497.04	497.08	0.01%
	Mean tax payment per vehicle	189.73	188.29	–0.76%
+ 20%	Mean annual VMT per vehicle	9684	9672	–0.12%
	Mean gas consumption per vehicle	442.26	442.30	0.01%
	Mean tax payment per vehicle	168.89	167.54	–0.80%
+ 50%	Mean annual VMT per vehicle	9167	9156	–0.12%
	Mean gas consumption per vehicle	418.78	418.82	0.01%
	Mean tax payment per vehicle	159.96	158.64	–0.83%
+ 100%	Mean annual VMT per vehicle	8307	8296	–0.14%
	Mean gas consumption per vehicle	379.65	379.69	0.01%
	Mean tax payment per vehicle	145.08	143.82	–0.87%

Note: VMT is measured by miles per year. Gasoline consumptions are measured by gallons. Tax payments are all in 2009 dollars.

Table 8. Changes in Annual VMT, Gasoline Consumption, and Tax Revenue under Optimal Gasoline Tax

Variable	Gas tax	Optimal gas tax	Percentage change
Mean annual VMT per vehicle	10028	9625	–4.015%
Mean gas consumption per vehicle	454.73	439.58	–3.332%
Mean tax burden per vehicle	173.46	443.98	155.946%

Note: VMT is measured by miles per year. Gasoline consumptions are all in gallons. Tax burdens are all in 2009 dollars.

Raising Gasoline Tax to the Optimal Level. Finally, we examine how driving demand and tax revenues would respond if governments raised the gasoline tax rates to $1.01 per gallon, which is the optimal fuel tax estimated by Parry and Small (2005). We replace all federal and state taxes with the $1.01 per gallon tax rate in all states and estimate the changes in the average annual VMT, gasoline consumption, and average tax payments per vehicle. Table 8 reports the simulation result at the national level. We find that under the new gasoline tax rates, total annual VMT would decrease by 4% and gasoline consumption would decrease by 3.3%. However, the tax payments would increase by more than 150%.

Table 9 reports our simulation results decomposed by income groups. All income groups would be affected severely should the optimal gasoline tax be adopted. In particular, the lower-income households would have to pay hundreds extra in tax money every year. Although the magnitude of increase in tax payment is somewhat uniform across households with different income levels, it imposes a much heavier burden on the lower-income households because their tax payments account for a larger share of their incomes. This finding suggests that substantially raising the gasoline tax rate could lead to equity concerns.

Conclusion

Over the past few years, increased attention has been paid to VMT tax as one possible solution to the shortfalls in government funding for transportation infrastructure. However, there has been little research investigating the behavioral response to a possible nationwide VMT tax and its fiscal and distributional implications. In this paper, we examine the viability of a revenue-neutral flat VMT tax using the 2009 NHTS data. We compute a revenue-neutral flat VMT tax rate for each state. We expect our parameters would be valuable for state policymakers who are considering and planning for pilot tests of mileage fee policy. Our simulation results confirm that a VMT tax could provide much more stable tax revenues than the gasoline tax, particularly when the nation's fleet fuel economy

Table 9. Tax Burden Change across Income Groups under the Optimal Gas Tax

U.S.($)	Mean annual household gas tax burden	Mean annual household optimal tax burden	Mean change	Percentage change
under 5,000	154.02	391.51	237.48	154%
5,000–9,999	135.31	341.76	206.45	153%
10,000–14,999	143.93	367.10	223.16	155%
15,000–19,999	167.15	428.52	261.37	156%
20,000–24,999	194.01	491.96	297.95	154%
25,000–29,999	221.70	563.96	342.26	154%
30,000–34,999	251.14	645.58	394.44	157%
35,000–39,999	273.11	695.60	422.49	155%
40,000–44,999	296.70	749.14	452.44	152%
45,000–49,999	310.95	796.83	485.88	156%
50,000–54,999	340.11	871.58	531.47	156%
55,000–59,999	355.65	907.84	552.19	155%
60,000–64,999	379.89	978.45	598.56	158%
65,000–69,999	385.13	988.31	603.18	157%
70,000–74,999	408.94	1042.57	633.63	155%
75,000–79,999	417.01	1069.14	652.12	156%
80,000–99,999	451.42	1163.26	711.84	158%
above 100,000	491.37	1260.73	769.36	157%

Note: Tax burdens are all in 2009 dollars.

continues to increase to meet the renewed CAFE standards. Regarding the distributional implication, our simulation suggests very small differences between a VMT tax and the current gasoline taxes. If the government raised the current gasoline tax substantially to account for the externalities and generate more revenues, it would impose an even larger burden on the lower-income groups. Overall, we conclude that a VMT tax could serve as a viable alternative to gasoline taxes.

One caveat with our findings is that our estimates are confined to gasoline tax and light-duty vehicles which constitute the majority of the vehicle models in the NHTS data used in our empirical analysis. Considering that diesel fuel taxes constitute a significant proportion of the total fuel tax revenues, a more comprehensive analysis of VMT tax should also include heavy-duty vehicles. Also, it is worth noting that in this study we assume governments adopt a flat VMT tax rate. Nonetheless, a VMT tax rate could also be designed to vary by vehicle type, class of road and congestion levels to address different policy issues. For example, policymakers could adopt a lower VMT tax rate for more fuel-efficient vehicles to incentivize fuel conservation. We leave studying the design of differential VMT tax rates for future research.

Finally, it should be noted that adopting a VMT tax may still raise other concerns, especially about the transition and administration costs. The transition to a VMT tax would incur a significant upfront cost of equipment installation. If the government keeps both the VMT tax and the gasoline tax in place during the transition period, the administrative cost of tax collection would be even higher. More research needs to be conducted to examine and understand the pattern of behavioral changes in response to the new VMT tax. It is also important to recognize that, despite its revenue-generating advantage, adopting a VMT tax faces limitations in terms of cost, administrative challenges, and user acceptance, and poses an array of technical, institutional, and political uncertainties (*20*, *21*). Thus, additional policy experiments and trials are needed to resolve these issues along with the research efforts.

Author Contributions

The authors confirm contribution to the paper as follows: study conception and design: YW; data collection: YW; analysis and interpretation of results: YW, QM; draft manuscript preparation: YW, QM. All authors reviewed the results and approved the final version of the manuscript.

References

1. National Surface Transportation Infrastructure Financing Commission. *Paying Our Way: A New Framework for Transportation Finance*. National Surface Transportation Infrastructure Financing Commission, Federal Advisory Committee to the U.S. Department of Transportation, Washington, D.C., 2009.
2. Whitty, J. M. *Oregon's Mileage Fee Concept and Road User Fee Pilot Program*. Final Report. Oregon Department of Transportation, Salem, OR, 2007.
3. Congressional Budget Office. *Alternative Approaches to Funding Highways*. Congressional Budget Office, Congress of the United States, Washington, D.C., 2011.

4. Schank, J., and N. Rudnick-Thorpe. End of the Highway Trust Fund? Long-Term Options for Funding Federal Surface Transportation. *Transportation Research Record: Journal of the Transportation Research Board*, 2011. 2221: 1–9.
5. Forkenbrock, D. J. Implementing a Mileage-Based Road User Charge. *Public Works Management Policy*, Vol. 10, No. 2, 2005, pp. 87–100.
6. Weatherford, B. A. Distributional Implications of Replacing the Federal Fuel Tax with Per Mile User Charges. *Transportation Research Record: Journal of the Transportation Research Board*, 2011. 2221: 19–26.
7. McMullen, B. S., L. Zhang, and K. Nakahara. Distributional Impacts of Changing from a Gasoline Tax to a Vehicle-mile Tax for Light Vehicles: A Case Study of Oregon. *Transport Policy*, Vol. 17, No. 6, 2010, pp. 359–366.
8. Robitaille, A., M. Jasmy, and L. Zhang. Effectiveness and Equity of Vehicle Mileage Fee at Federal and State Levels. *Transportation Research Record: Journal of the Transportation Research Board*, 2011. 2221: 27–38.
9. Greene, D. L. What is Greener than a VMT Tax? The Case for an Indexed Energy User Fee to Finance Us Surface Transportation. *Transportation Research Part D: Transport and Environment*, Vol. 16, No. 6, 2011, pp. 451–458.
10. Pickrell, D., and P. Schimek. Growth in Motor Vehicle Ownership and Use: Evidence from the Nationwide Personal Transportation Survey. *Journal of Transportation and Statistics*, Vol. 2, No. 1, 1999, pp. 1–17.
11. Schimek, P. Gasoline and Travel Demand Models Using Time Series and Cross-Section Data from the United States. *Transportation Research Record: Journal of the Transportation Research Board*, 1996. 1558: 83–89.
12. Parry, I. W., and K. A. Small. Does Britain or the United States Have the Right Gasoline Tax? *American Economic Review*, Vol. 95, No. 4, 2005, pp. 1276–1289.
13. Hanley, P. F., and J. G. Kuhl. National Evaluation of Mileage-Based Charges for Drivers, Initial Results. *Transportation Research Record: Journal of the Transportation Research Board*, 2011. 2221: 10–18.
14. Rufolo, A. Cost Estimates for Collecting Fees for Vehicle Miles Traveled. *Transportation Research Record: Journal of the Transportation Research Board*, 2011. 2221: 39–45.
15. Agrawal, A. W., H. Nixon, and A. M. Hooper. *NCHRP Synthesis 487: Public Perception of Mileage-based User Fees*. Transportation Research Board of the National Academies, Washington, D.C., 2016.
16. Zhang, L., B. S. McMullen, D. Valluri, and K. Nakahara. The Short- and Long-Run Impact of a Mileage Fee on Income and Spatial Equity. *Transportation Research Record: Journal of the Transportation Research Board*, 2009. 2115: 110–118.
17. Caplan, A. J. Estimating the Effectiveness of a Vehicle Miles Traveled Tax in Reducing Particulate Matter Emissions. *Journal of Environmental Planning and Management*, Vol. 52, No. 3, 2009, pp. 315–344.
18. Greene, D. L. Rebound 2007: Analysis of U.S. Light-duty Vehicle Travel Statistics. *Energy Policy*, Vol. 41, 2012, pp. 14–28.
19. Sorrell, S., and J. Dimitropoulos. The Rebound Effect: Microeconomic Definitions, Limitations and Extensions. *Ecological Economics*, Vol. 65, 2008, pp. 636–649.
20. Sorensen, P., M. Waches, and L. Ecola. *NCHRP Web-Only Document 161: System Trials to Demonstrate Mileage-Based Road Use Charges*. Transportation Research Board of the National Academies, Washington, D.C., 2010.
21. Sorensen, P., L. Ecola, M. Wachs, M. Donath, L. Munnich, and B. Serian. *NCHRP Web-Only Document 143: Implementable Strategies for Shifting to Direct Usage-Based Charges for Transportation Funding*. Transportation Research Board of the National Academies, Washington, D.C., 2009.

The Standing Committee on Revenue and Finance (ABE10) peer-reviewed this paper (18-04245).

Article

TRR
JOURNAL OF THE TRANSPORTATION RESEARCH BOARD

Transportation Research Record
2018, Vol. 2672(4) 43–53
© National Academy of Sciences:
Transportation Research Board 2018
Article reuse guidelines:
sagepub.com/journals-permissions
DOI: 10.1177/0361198118791629
journals.sagepub.com/home/trr
SAGE

Evaluating the Effectiveness of Public–Private Partnerships in Highway Development: The Case of Virginia

Yin Wang[1] and Zhirong Jerry Zhao[2,3]

Abstract

Given the current momentum for public–private partnerships (PPPs), it is critical to review the experiences of PPP highway projects to see whether they succeed in serving public benefits. This article applies a goal-centered approach to evaluate the effectiveness of nine PPP highway projects in the Commonwealth of Virginia, U.S.A., that were implemented and opened to traffic between 1990 and 2016. Virginia has used highway PPPs more for financing or risk reduction than for efficiency gains. The authors examine four elements of contract agreements—PPP type, private partner selection, financial arrangements, and risk allocation—in these Virginian projects, and find that these arrangements have been effective in accessing innovative finance and preventing cost overrun, while the evidence is limited regarding shifting revenue risk or achieving efficiency gains.

Public–private partnerships (PPPs) have been adopted globally for infrastructure development in the past few decades. But the use of PPP has remained low in the transportation sector across the United States, probably due to a lack of fiscal incentives associated with the traditional funding mechanism. The highway system in the U.S.A. is funded mainly with dedicated funding sources, such as federal and state highway trust funds with earmarked fuel taxes. In addition, state and local governments can issue municipal bonds, which, compared with private financing, tend to have lower borrowing costs because their interests are exempted from federal income tax. In recent years, many states have begun to explore PPP in state highway development, due to recurring budget crises and the shortage of transportation funding. The path to PPP, however, is not smooth. The confusion and controversy surrounding recent concession cases in the U.S.A. has led to widespread public concern and legislative caution, in particular, on whether partnership arrangements between the public and private sectors can effectively advance the public interest (*1*).

As of 2016, 35 U.S. states, the District of Columbia, and one U.S. territory had enacted legislation enabling the use of PPP (*2*). Geddes and Wagner (*3*) examine why states pass PPP-enabling laws and the features of PPP-enabling laws. Among other things, they find that larger transportation demand, indicated by vehicle registration growth and greater traffic congestion, enhances the likelihood of passage, and a higher percentage of Republicans in the state's House of Representatives increases a law's favorability to private investment. The wide adoption of enabling laws signifies the increasing interest in PPP, but the actual use of PPP is far more limited. According to the Federal Highway Administration (FHWA), by 2016, a total of 78 transportation projects had been developed under various PPP models, such as Design–Build–Finance–Operate–Maintain (DBFOM) and long-term lease (LTL) concession (*2*). These projects were highly concentrated in four states: Virginia, Texas, Florida, and Colorado, which together built 35.

Given the current momentum for consideration of PPP, it is critical to examine the experiences of these highway PPP projects and to evaluate their effectiveness in serving public benefits. The literature on PPPs is filled with either ideological advocacy or criticism, but remains thin on systematic and rigorous evaluation of PPP performance (*4*), and none has been conducted from the perspective of a state in the U.S. context. In this study, the authors develop a goal-centered approach to evaluate

[1]School of Public Economics and Administration, Shanghai University of Finance and Economics, Shanghai, China
[2]Hubert H. Humphrey School of Public Affairs, University of Minnesota, Minneapolis, MN
[3]China Institute for Urban Governance, Shanghai Jiaotong University, China

Corresponding Author:
Address correspondence to Zhirong Jerry Zhao: zrzhao@umn.edu

Table 1. PPP in the Spectrum of Project Delivery Options

Project delivery taxonomy		The full spectrum of project delivery						
		Traditional delivery	Public–private partnerships (PPP)				Full privatization	
			Public-finance PPP		Private-finance PPP			
Facility development stage	Design/ Construction	Design–Bid–Build (DBB)	Design–Build (DB)	Design–Build–Operate–Maintain (DBOM)	Design–Build–Finance(DBF)	Design–Build–Finance–Operate–Maintain (DBFOM)		Build-Own-Operate (BOO)
	Operation/ Maintenance	Public Operate			Long–Term Lease (LTL)		Asset Sale	
	Expansion/ Rehabilitation	DBB + Public Operate	Design–Build–Operate–Maintain (DBOM)		Lease–Build–Operate (LBO); Wraparound Addition			Buy–Build–Operate (BBO)

Source: (*1*).

the effectiveness of PPP by comparing project outcomes with the goals of governments in initiating the partnerships. This framework is empirically applied to a multiple case study of PPP highway projects in Virginia that were implemented and open to traffic between 1990 and 2016. The research seeks to answer three questions: (a) What does Virginia aim to accomplish with highway PPP? (b) What do the agreements look like in these PPP contracts? (c) Are these PPP contract agreements effective in achieving the intended project outcomes?

This article is organized as follows. The next section reviews the literature on the concept of PPP, government rationales for PPP, and empirical assessments of PPP effectiveness. The third section discusses the goal-centered framework of PPP evaluation. The fourth and fifth sections explain the case selection and research methodology, and then present the analysis of Virginia's highway PPP. The final section concludes and addresses policy implications and research recommendations.

Literature Review

The literature on PPPs is filled with either ideological advocacy or criticism. Systematic and rigorous evaluation of PPP performance, however, is still in its nascent stage. Challenges come from the ambiguity of the concept of PPP, the breadth of PPP goals and objectives, and limited evidence of the effectiveness of PPPs.

The Concept of Highway PPP in the U.S. Context

The term "public–private partnerships" (PPP) is defined in many different ways, coming from a rich history of public–private mixing across multiple lineages. Hodge and Greve (*5*) argue that the PPP semantics have been used internationally across five different families of arrangements: (a) institutional cooperation for joint production and risk sharing; (b) long-term infrastructure contracts (LTIC); (c) public policy networks; (d) civil society and community development; and (e) urban renewal and downtown economic development. This study focuses on PPP in highway development in the U.S.A., which is akin to Hodge and Greve's notion of LTIC-type PPP but may also include some shorter-term arrangements. The U.S. Department of Transportation (USDOT) defines PPP as formal contractual agreements between public and private sector partners that "allow more private sector participation than is traditional" (*6*) (p. 10). This definition includes infrastructure delivery types either with the involvement of private finance, similar to UK private finance initiatives (PFI), or without private finance, such as Design–Build (DB).

To provide a taxonomy of PPP in the U.S.A., Zhao (*1*) categorizes various PPP options together with other modes of infrastructure development along two dimensions: facility development stage, and level and type of private involvement (Table 1). Vertically, project delivery options differ with respect to stages of facility development, which include design and construction, operation and maintenance, and expansion or rehabilitation. Horizontally, the taxonomy shows the spectrum of public–private involvement in project delivery. Between the two extremes of traditional project delivery and full privatization, are public-finance PPP and private-finance PPP options.

Typical examples of public-finance PPP include Design–Build (DB) and Design–Build–Operate–Maintain (DBOM). DB differs from the traditional approach of Design–Bid–Build (DBB) by combining design and construction phases into a single contract, aiming for improved efficiency. In DBOM, DB is further combined with the stage of operation and maintenance, in which the private sector is compensated after meeting specific performance requirements. In contrast, private-finance PPP options involve a certain percentage of private finance, and thus contain an "F" component, such as Design–Build–Finance (DBF) and Design–Build–Finance–Operate–Maintain (DBFOM). An additional example is LTL, in which an existing public facility is leased to a private sector concessionaire that pays an upfront amount to the government and then receives returns by operating the facility.

Multiplicity of PPP Objectives and Evaluation Frameworks

The objectives in use of PPP may vary from place to place and change over time. The UK's PFI was initiated in the early 1990s, with the primary rationale being to get around restrictions on formal public-sector debt levels (*7*). This was followed by the goal of reducing pressure on public sector budgets. PFI was later further justified with the promise to provide better value for money (VfM) for taxpayers. Besides these three initial promises, Hodge (*7*) lists 10 more, including some governance values such as better accountability and greater innovation.

Most evaluations of PPP focus on a few performance dimensions that are emphasized from different disciplinary perspectives. Economists emphasize evaluations of relative managerial efficiency and allocative efficiency across different infrastructure delivery methods. Political scientists focus on the arena of PPP-related policy debates and multiple types of governance. Policy and planning scholars emphasize the extent to which policy goals are met amidst negotiated agreements, the need for both procedural and analytical rationality, and the numerous conflicts of interest buried inside PPP adoption and implementation (*7*). In this study, the authors follow the policy and planning approach to evaluate the effectiveness of PPP adoption in achieving policy goals.

The Limit of VfM Appraisal in Justifying PPP Effectiveness

LTIC-type PPP is often justified by VfM appraisal, an ex-ante calculation to assess the extent to which cost savings could be achieved through PPP relative to a traditional delivery scheme. The fundamental validity and legitimacy of VfM appraisal, however, is under question (*8*). Siemiatycki and Farooqi (*9*) examine 28 VfM reports made by the government of Ontario, Canada. The average costs of these projects were shown to be 11% lower when delivered through PPP than through traditional public procurement, but the positive results were driven by the transfer of unusually high risks from government to the private partners. The calculation seems to be biased toward the use of PPP, raising concerns about several governance issues, especially the lack of transparency in the PPP planning stage. An additional issue with VfM is that the outcomes are likely to change during the course of PPP implementation after the ex-ante VfM analysis. Severe cost overruns compared with the original contracts have been seen in many cases (*5*). Moreover, under the current budgetary constraints in the public sector worldwide, many PPP projects are motivated by the promise of additional financing from the private sector. The realistic choice is probably not between a PPP and public-sector procurement of a facility, but between a PPP and no investment at all.

A Goal-Centered Framework for PPP Evaluation

Following Dahler-Larsen's (*10*) rationale that the goals of a program constitute the primary source of criteria for judging the program, this study proposes a goal-centered framework for PPP evaluation, which assesses the effectiveness of PPP arrangements by comparing PPP project outcomes against particular goals. The framework is illustrated in Figure 1. The oval space represents the

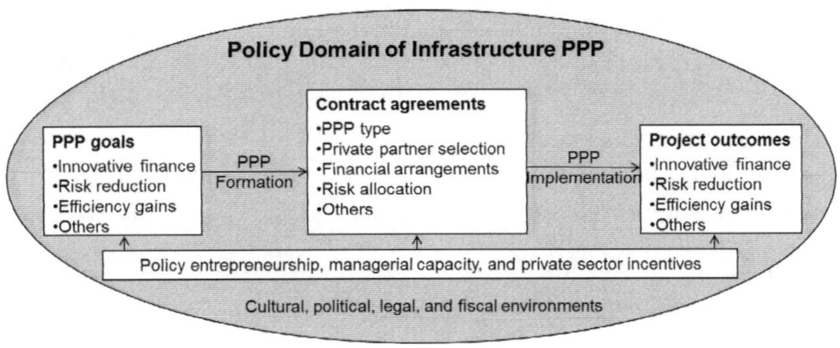

Figure 1. A goal-centered framework for PPP evaluation.

infrastructure PPP policy domain. In the center of the policy domain is the pathway of PPP performance, in which partnership contract agreements are the link between PPP goals and project outcomes. PPP contracts are used as a control tool to help the government protect its own goals (*11*), as detailed agreements contained in the contracts would guide the collaborative activities in the implementation of PPP. The framework also includes a set of human factors at the governance level (underneath the pathway) and the background of policy-level contextual variants (surrounding the pathway), but the focus of this study is the PPP pathway with three key components: PPP goals, contract agreements, and project outcomes.

Frequently cited goals for adopting PPP include innovative finance, reduced risks, and efficiency gains. PPP may help the government bypass debt limits or procedure constraints associated with governmental borrowing, or reduce government's immediate expenditures to keep budget deficits down (*12*). PPP may be designed to shift the risks of cost overrun, schedule delay, or revenue shortfall to the private sector (*13*). Moreover, PPP arrangements may be more efficient in project delivery compared with traditional public procurement. When a private partner takes a life-cycle approach to project development, the integration of multiple project phases may result in accelerated delivery and cost saving.

PPP contract agreements have four key elements: PPP type, private partner selection, financial arrangements, and risk allocation. The choice of PPP type mainly concerns the stage of facility development to involve private partners—DB, operation and maintenance, or the combination of them—and whether private partners would contribute a proportion of initial finance. Private partner selection may involve specific criteria, restrictions, or procedure requirements. The selected private partner must be viewed as trustworthy by government, with positive expectations of both the partner's intentions and its competence (*14*). Regarding financial arrangements, private-finance PPP brings additional finance such as private equity and private debt for infrastructure projects. Lastly, how risks are allocated in contract agreements is also critical to PPP performance. There are "elemental risks" that are associated with specific stages of facility development—land acquisition or environmental clearance issues, for instance, are typically faced in the planning stage; and "global risks" that could occur at any project stage, such as uncertainty associated with political environment, financial market, or other unforeseeable and uncontrollable force majeure (*1*).

Contract agreements reached between government and the private sector govern the implementation process of PPP. The results of the implementation process are project outcomes, which this research intends to evaluate. Scrutiny of such targeted performance is especially important for infrastructure PPP in the U.S.A., where the adoption of PPP is still in its nascent stage and the efficacy of it to advance public value remains contested.

Case Selection and Research Methodology

This study empirically applies the goal-centered evaluation framework to Virginia and evaluates the effectiveness of PPP highway projects there that were implemented and open to traffic between 1990 and 2016. Virginia has been one of the states most active in PPP highway development in the U.S.A. since the end of the Interstate Highway Era in the 1980s (*15*). The Virginia General Assembly passed the Highway Corporation Act (HCA) in 1988, which authorized Dulles Greenway, the state's first PPP project. The law delegated to the Virginia State Corporation Commission (VSCC) the powers to regulate this project like a public utility. In 1995, the General Assembly authorized the Virginia Public-Private Transportation Act (PPTA), under which the Virginia Department of Transportation (VDOT) is responsible for PPP road development. PPTA was further modified several times with later amendments. Overall, across its 30-year history, Virginia's PPP statutory framework has evolved substantially in the directions of increasing program flexibility and addressing public interest concerns (*16*). Arguably, this legislative development has contributed significantly to Virginia's active and successful experience with PPP.

As of 2016, Virginia had developed nine PPP highway projects that were open to traffic, tied with Florida, more than any other state in the U.S.A. (*2*). By evaluating these nine Virginian projects, this study will not only shed light on an important part of U.S. PPP experiences, but also have practical implications for other states that are still debating about PPP as a policy option for innovative project delivery.

The use of a comparative case study approach is necessitated here by the multiplicity of partnership goals to be evaluated, the multiplicity of contract agreements believed to influence project performance, and the limited number of PPP projects in the state. Detailed information on the projects was collected from multiple sources, including Virginia P3 Office, the Library of Virginia, and the Innovative Program Delivery Office at FHWA. The authors also referred extensively to local media coverage, which pays close attention to the development of local projects. The authors searched the websites of 37 organizations consisting of local newspapers, television stations, radio stations, and construction industry, and retrieved more than 300 related articles and reports. Lastly, the authors interviewed three transportation consultants, who have extensive experience of PPP development in the U.S.A. The interviews were semi-structured, with a set of guide questions and various follow-up questions. For data analysis, the authors adopted pattern-matching techniques (*17*) for cross-case analysis and utilized process tracing techniques (*18*) for within-case analysis. The rich qualitative data collected on the Virginia cases enabled the use of these research approaches, which require detailed information on contexts, processes, and outcomes.

Table 2 lists the nine projects, numbered by the date when their PPP contracts were signed. To facilitate the exploration of case patterns, these projects are coded with three project attributes: whether tolling is used ("T" for toll, "N" for non-toll), whether private finance is involved ("P" for private finance, "U" for no private finance), and project type ("G" for greenfield, "B" for brownfield, and "H" for high-occupancy toll, HOT). The table also includes information about PPP type, project cost, concession term (if applicable), and project status, and a short description of each project. The nine cases include the most common PPP types such as DB, DBF, DBFOM, and LTL).

Evaluating the Effectiveness of Virginia's Highway PPPs

What Does Virginia Aim to Accomplish with Highway PPPs?

The authors coded specific objectives of the projects that were explicitly expressed in the collected documents.

Table 2. List of PPP Highway Projects in Virginia

ID	Year	Coding	Project name	PPP type	Cost ($ million)	Concession term (yrs)	Status
1	1993	1-TPG	Dulles Greenway	DBFOM	$350	42.5	In operation
	Project description: a 14-mile toll road, developed by a consortium led by a wealthy family in Virginia, and financed with full private finance.						
2	1998	2-TPG	Pocahontas Parkway	DBFOM	$381	30	Early termination
	Project description: an 8.8-mile toll road, developed by a corporation created by VDOT, and financed largely with non-profit tax-exempt bonds.						
3	2000	3-NPG	Route 288	DBF	$236	–	In operation
	Project description: a 10.5-mile toll-free portion of Route 288, developed by a consortium led by a Virginian construction company, financed with bonds issued by the private developer backed by state transportation trust funds.						
4	2002	4-NUB	Route 199	DB	$32	–	In operation
	Project description: a stretch of 5-mile road improvements of the toll-free Route 199, built by a consortium consisting of three Virginian construction companies, and funded almost entirely with traditional state finance.						
5	2002	5-NUB	Route 28 Corridor Improvements	DB	$321	–	Partially completed
	Project description: a stretch of 14-mile road improvements to Route 28 Corridor, built by a consortium led by a Virginian construction company, and financed with government grants and state transportation bonds.						
6	2003	6-NUB	U.S. Route 58 Corridor Development	DB	$223	–	Partially completed
	Project description: a stretch of 36-mile road improvements of Route 58 Corridor, built by a Virginian highway construction company, and financed with state General Obligation Bonds backed by state transportation trust fund.						
7	2006	7-TPB	Pocahontas Parkway Lease	LTL	$666	99	In operation
	Project description: LTL for the existing Pocahontas Parkway (2-TPG), taken over by Transurban, and financed with private equity, private debt, and TIFIA loans.						
8	2006	8-TPH	I-95 Express Lanes	DBFOM	$916	76	In operation
	Project description: 29.4-mile HOT lanes along I-95 and I-395 corridor, developed by a joint venture of Transurban and Fluor, and financed with private equity, TIFIA loans, PABs, and public grants.						
9	2007	9-TPH	Capital Beltway (I-495) Express Lanes	DBFOM	$2,068	85	In operation
	Project description: 14-mile HOT lanes on the Capital Beltway, developed by a joint venture of Transurban and Fluor, and financed with private equity, TIFIA loans, PABs, and public grants.						

Coding Rule: #-XYZ

#	Project ID, numbered by date of the PPP contract signed.
X	T: The project is tolled; N: The project is not tolled.
Y	P: It involves private financing; U: It does not involve private financing.
Z	G: Greenfield project; B: Brownfield project; H: HOT lanes project.

Note: – = not applicable.

Table 3. Goals and Objectives of Virginia's Highway PPPs

Goals	Objectives	Project	PPP type	P*
Innovative finance	Accelerated financing	1-TPG	DBFOM	✔
		2-TPG	DBFOM	✔
		9-TPH	DBFOM	✔
	Additional financing	3-NPG	DBF	✔
		8-TPH	DBFOM	✔
	Revenue contribution	7-TPB	LTL	✔
		8-TPH	DBFOM	
Risk reduction	Shifting revenue risk	1-TPG	DBFOM	
		2-TPG	DBFOM	
		7-TPB	LTL	
		8-TPH	DBFOM	
		9-TPH	DBFOM	
	Preventing cost overrun	1-TPG	DBFOM	
		2-TPG	DBFOM	
		3-NPG	DBF	
		4-NUB	DB	✔
		5-NUB	DB	✔
		6-NUB	DB	✔
		8-TPH	DBFOM	
		9-TPH	DBFOM	
	Preventing construction delay	1-TPG	DBFOM	
		2-TPG	DBFOM	
		3-NPG	DBF	
		4-NUB	DB	✔
		5-NUB	DB	✔
		6-NUB	DB	✔
		8-TPH	DBFOM	
		9-TPH	DBFOM	
Efficiency gains	Cost saving	3-NPG	DBF	
	Time saving	3-NPG	DBF	

Note: * = as the primary objective expressed in the project.

These objectives fall neatly into three categories of project-level goals: innovative finance, reduction of risks to the government, and efficiency gains (Table 3). Meanwhile, the primary objective of each project was also identified, as denoted in the last column in Table 3.

Firstly, accessing innovative finance is a goal expressed for six projects built under PPP models with a financing component: DBFOM, DBF, and LTL. Three projects (1-TPG, 2-TPG, and 9-TPH) used PPP for "accelerated financing," in which private financial resources would be used to develop projects that otherwise would have to wait for later deployment. Two projects (3-NPG and 8-TPH) adopted PPP for "additional financing," in which private financial resources would be used for projects that otherwise would be smaller in scale or receive no investment at all. Two projects (7-TPB and 8-TPH) utilized PPP for "revenue contribution," in which a lump-sum payment would be paid by a private entity to the government in exchange for operation rights of a facility.

Secondly, all of the nine projects adopted PPP with a goal of risk reduction, which pertains to various specific objectives across the cases. The five cases with tolls expressed an objective of "shifting revenue risk," which is to transfer the risk of toll revenue shortfall from government to the private sector. Meanwhile, eight of the projects with a "DB" component explicitly expressed the objectives of "preventing cost overrun" and "preventing construction delay," aiming at completing the projects within the estimated cost and completion date specified in PPP contracts. Thirdly, only one project in Virginia, 3-NPG, cited efficiency gains as a reason for adopting the PPP approach. Here, efficiency gains refer to both "cost saving" and "time saving" that may be achieved under PPP compared with traditional public procurement.

The authors also coded the primary objective for each project, denoted in the last column in Table 3, based on two criteria: (a) whether a particular goal is clearly identified as the first and most important one in governmental documentation, and (b) whether a particular goal is consistently covered in multiple governmental documentation and media stories regarding the case. A pattern-matching analysis shows that all six projects with a financing component adopted PPP primarily for a finance purpose, and that the other three projects (all DB cases) adopted PPP primarily for risk reduction and for preventing cost overrun and construction delay. In short, Virginia has used PPP primarily for financing or risk reduction, but much less for efficiency gains.

What Do the Agreements Look Like in These PPP Contracts?

The authors go on to examine the features of PPP agreements that were reached in these Virginian PPP contracts. Firstly, regarding PPP type, it is worth noting that not all PPP projects involve private finance, tolling, or concession. Three (4-NUB, 5-NUB, and 6-NUB) among the nine projects involved only a DB contract, in which the private contractor assumed design and construction roles but did not provide private finance. This is a distinctive feature of PPP in the U.S.A., unlike PFI in the UK which always involves private finance. In another project (3-NPG), while financed with bonds issued by the private developer, the facility was not tolled and involved no concession. The borrowing was paid back by state transportation trust funds.

Secondly, two observations were made on the selection of private partners. For toll concessions, the government's selection of private partners has become increasingly professional in the sector of toll road development. The two earliest toll roads built in the 1990s (1-TPG and 2-TPG) were developed by domestic corporations with less expertise in highway tolling. In the three recent toll projects built in the 2000s (7-TPB, 8-TPH and 9-TPH), the government chose to partner with multinational professional toll road developers such as Transurban, which have strong financial and managerial capabilities related to toll road development and operation. Moreover, the state government has chosen to work with the same partner for multiple projects, which is an indicator of a mature collaborative relationship. Fluor led a team to construct 2-TPG in the 1990s and later teamed up with Transurban to develop both 8-TPH and 9-TPH in the 2000s.

Thirdly, the financial arrangements of the nine projects are summarized in Table 4. Three DB projects (4-NUB, 5-NUB, and 6-NUB) are public-finance PPP, which were built with traditional public finance such as public grants and public bonds. The other six cases are private-finance PPP, with different levels of private finance involvement. The first and foremost observation is that there is a gradual shift from primary private finance to partnership finance, with a combination of public finance, private finance, and publicly-assisted finance. The percentage of private finance purely from market operation, including private debt financing and private equity, decreased gradually from 100% for the earliest project in the 1990s to less than 20% for the latest project in the late 2000s. The gap was mainly taken up by innovative government financing tools, including state infrastructure banks (SIB), Transportation Infrastructure Finance and Innovation Act (TIFIA) loans, and private-activity bonds (PABs). These tools may be called publicly-assisted finance. They come with more favorable interest rates than those in private capital markets (*19*), and thus may reduce the risk of private investment and enhance the financial attractiveness of PPP projects to the private sector. Take 7-TPB as an example. The TIFIA loan refinanced approximately $95 million of the long-term senior bank debt and paid about $55 million for supplementary construction costs. The size of the TIFIA support was determined through a cost-benefit analysis to incentivize the private partner to assume the risk of a project that would not have been economically feasible otherwise (*20*).

Another observation on the six projects utilizing private finance is that the involvement of private equity is more visible in the three latest cases. Private equity incurs a high level of financial risk and also requires a high rate of return for the private partner. A larger amount of private equity suggests that the government was able to get the private sector involved in a deeper way in financing PPP projects. The last observation on PPP financial arrangements is that the PPP projects with tolls received longer concession terms over time. The two earliest toll concessions—Dulles Greenway (1-TPG) and Pocahontas Parkway (2-TPG)—had concession terms of 42.5 and 30 years. After opening to traffic, their toll revenues were significantly lower than initial forecasts. As a result, the Virginian government extended the concession term of Dulles Greenway (1-TPG) for an additional 20 years and restructured Pocahontas Parkway (2-TPG) into a 99-year lease to keep the projects viable. Probably due to such early experiences, the later toll concessions (7-TPB, 8-TPH, and 9-TPH) all have much longer concession terms (76–99 years) compared with the earlier ones.

The fourth element of contract agreements examined in this study is risk allocation. Elemental risks associated with different stages of project development—administrative or regulatory risk, construction-related risk, and revenue risk—are allocated to either the government or the private sector per PPP contract (Table 4). Administrative or regulatory risk mainly concerns environmental permission and land acquisition. Across the Virginian PPP projects, the allocation of this risk has been adjusted over time. In the two earliest projects (1-TPG and 2-TPG), the responsibility for acquiring environmental clearance resided in the private sector, but the administrative processes turned out to be time consuming and costly. The Virginian government retained the responsibility in the public sector for more recent

Table 4. Financial Arrangements and Risk Allocation of Virginia's Highway PPPs

	1-TPG	2-TPG	3-NPG	4-NUB	5-NUB	6-NUB	7-TPB	8-TPH	9-TPH
PPP type	DBFOM	DBFOM	DBF	DB	DB	DB	LTL	DBFOM	DBFOM
Financial arrangements									
Project cost ($ million)	$350	$381	$236	$32	$321	$223	$666	$916	$2,068
Private finance (as percentage)	100%	93%	100%	0	0	0	78%	31%	17%
Debt service	Toll	Toll	Other	——	Other	Other	Toll	Toll	Toll
Concession Term (years)	42.5	30	——	——	——	——	99	76	85
Capital structure details									
Traditional public finance ($ million)	0	$9	0	$32	$321	$223	0	$83	$542
From federal	0	$9	0	0	0	0	0	0	0
From state	0	0	0	$32	$105	$223	0	$83	$495
From local	0	0	0	0	$180	0	0	0	0
Other	0	0	0	0	$36	0	0	0	$47
Private finance ($ million)	$350	$354	$236	0	0	0	$516	$280	$348
Private equity	$40	0	0	0	0	0	$141	$280	$348
Private debt	$310	$354	$236	0	0	0	$375	0	0
Innovative government financing tools ($ million)	0	$18	0	0	0	0	$150	$553	$1,178
SIB	0	$18	0	0	0	0	0	0	0
TIFIA	0	0	0	0	0	0	$150	$300	$589
PABs	0	0	0	0	0	0	0	$253	$589
Private risk									
Administrative/regulatory risk	√	√							
Construction (cost, time) risk	√	√	√	√	√	√	——	√	√
Revenue risk			——	——	——	——	√	√	√

Note: √ = risk allocated to the private sector; —— = not applicable.

projects. Construction-related risks are related to the possibility of cost overrun and construction delay, which often happens for mega projects (*13*). Across these projects, construction-related risks were always allocated to private partners, who were better able to manage these risks as they took the responsibility for design and construction of the road facilities.

Revenue risk pertains to the uncertainty associated with transportation demand and toll revenues. The two earliest toll projects were either regulated as a public utility by a government agency (1-TPG) or developed by a government-sponsored corporation (2-TPG); thus the ultimate revenue risk resided in the government. In 1-TPG, the authorizing legislation, the 1988 HCA, delegated authority to VSCC to regulate the road like a public utility (*21*), for which the state takes the ultimate responsibility and risk. In 2-TPG, the project developer was a corporation created by VDOT for the limited purpose of financing and developing the road. The event of the developer's inability to pay debt services would inevitably damage the state's overall financial creditworthiness (*22*). Therefore, the developer could not behave as an independent legal entity in the market to bear the revenue risk. In contrast, in the recent toll concessions (7-TPB, 8-TPH, and 9-TPH), under the authorization of the 1995 PPTA, Virginia changed its strategies and allocated the full risk to the private partners, which rely on project-generated toll revenues for investment returns.

Are These PPP Contract Agreements Effective in Achieving Intended Project Outcomes?

As illustrated in Table 5, PPP contract agreements have been largely effective in accessing innovative finance and preventing cost overrun, but they have achieved mixed results regarding the objectives of shifting revenue risk and achieving efficiency gains.

Firstly, reaching financial closure in private-finance PPP arrangements signifies the achievement of the finance goal, unless the implementation of project development is interrupted by other reasons. As expected, three projects (1-TPG, 2-TPG and 9-TPH) received private funding to accelerate project development. For example, the Pocahontas Parkway (2-TPG), which had been delayed for many years waiting for traditional public financing, received immediate financing from the private partner after signing the partnership contract in 1998. First planned in 1980, this project received tentative approval from FHWA in 1983 to be funded with Interstate 4R funds (4R stands for reconstruction,

Table 5. Project Outcomes of Virginia's Highway PPPs

	Project	Outcome	Notes
Innovative finance			
Accelerated financing	1-TPG	✔	Faster development with private equity and private debt
	2-TPG	✔	Faster development with financing from non-profit bonds
	9-TPH	✔	Faster delivery using private equity
Additional financing	3-NPG	✔	Larger financing from bonds issued by the private developer
	8-TPH	✔	Initiated with financing from private equity
Revenue contribution	7-TPB	✔	Upfront payment in private equity and private debt to VDOT
	8-TPH	×	No upfront payment due to changes in project design
Risk reduction			
Shifting revenue risk	1-TPG	×	Govt extended concession by 20 years for revenue shortfalls
	2-TPG	×	Govt terminated project and restructured it into 7-TPB
	7-TPB	✔	Bankruptcy with losses born by the private sector
	8-TPH	✔	Performed well financially since opening in Dec. 2014
	9-TPH	✔	Lost over $300 million by now, borne by the private sector
Preventing cost overrun	1-TPG	✔	Under-budget delivery
	2-TPG	✔	Completed $10 m below the project's $324 m contract
	3-NPG	✔	Cost $236 m as agreed in the contract
	4-NUB	✔	Completed on-budget
	5-NUB	✔	So far all project portions have been delivered on budget
	6-NUB	✔	So far all project portions have been constructed on budget
	8-TPH	✔	Completed on-budget
	9-TPH	✔	Completed on-budget
Preventing construction delay	1-TPG	✔	Opened to traffic 6 months ahead of schedule
	2-TPG	×	Completed 5 months later than the target completion date
	3-NPG	×	Record levels of rainfall caused a one-year delay
	4-NUB	✔	Completed 71 days ahead of schedule
	5-NUB	✔	Project portions have been completed ahead of schedule
	6-NUB	✔	So far project portions have been constructed on time
	8-TPH	✔	Opened to traffic on time
	9-TPH	✔	Opened a month ahead of schedule
Efficiency gains			
Cost saving	3-NPG	✔	Project cost $236 million, compared with public estimate of $283 m
Time saving	3-NPG	×	Longer delivery than VDOT's estimate due to high rainfall

Note: ✔ = Goal fulfilled; × = Goal unfulfilled.

restoration, rehabilitation, and resurfacing), which did not become available. Without the private finance, the road could not have been constructed for at least another 15 years, according to VDOT's forecasts in 1998 (*23*).

Two other projects (3-NPG and 8-TPH) raised private finance to build road facilities that otherwise may have been smaller in scale or received no investment at all. For instance, Route 288 (3-NPG) received larger-scale financing raised through private bonds, which allowed the project to be built as an Interstate-standard four-lane freeway, instead of a mostly two-lane state route as originally planned with VDOT funding (*24*). In addition, the Pocahontas Parkway Lease (7-TPB) generated an upfront payment from the private developer to the public sector, as Transurban paid a total amount of $611 million to acquire the rights to operate the road for 99 years (*20*). In I-95 Express Lanes (8-TPH), however, revenue contribution was not realized as planned. Due to an environmental lawsuit filed by a local government, the state government changed the project design and shortened the road, which significantly lessened the project's net toll revenues, and accordingly removed the revenue contribution requirement.

Secondly, the performance on risk reduction has been mixed. Regarding the objective of shifting revenue risk, earlier toll projects did not achieve this, while recent toll projects tended to perform better. In the two earliest projects (1-TPG and 2-TPG), the government was unable to fully shift the revenue risk to the private sector and had to intervene with a rescue plan in the event of severe revenue shortfalls. For 1-TPG, VSCC extended the concession period for an additional 20 years in order to increase the project's financial viability. For 2-TPG, VDOT terminated the franchise ahead of schedule and restructured the project into a new PPP, 7-TPB. In contrast, in two recent projects (7-TPB and 9-TPH), the government effectively transferred the revenue risk to the private sector. When Pocahontas Parkway Lease (7-

TPB) went bankrupt in June 2013, the private partner wrote off the asset from its balance sheet and transferred control of the highway to its creditor. The private sector bore substantial losses while this transition neither exposed the Virginian government to any financial loss under the partnership agreement nor affected service for road users (*25*). Notably, the federal government also took some loss: TIFIA Credit Program sold its interest in the project, a $150 million loan, to some private companies for approximately $60 million (*20*). Project 9-TPH lost more than $300 million between its opening in November 2012 and December 2016. The full losses were absorbed by the private firm that developed and operates the facility.

Regarding the objective of preventing cost overrun, the analysis shows all of the eight projects intended to do so were delivered on budget. These results are remarkable because cost overruns are very common among mega projects. Among the same eight projects, six were successful in preventing construction delay. Project 2-TPG was completed five months later than the target completion date, resulting in a payment of $1.9 million in liquidated damages from the design–build contractor to compensate the loss of value for the public. Project 3-NPG was delayed by one year by record levels of rainfall; the private contractor shouldered the direct cost of the construction delay.

Lastly, there is a lack of evidence of efficiency gains in Virginia's PPP projects. Besides the fact that only one project, 3-NPG, explicitly expressed a PPP goal of efficiency gains, the project did not realize "time saving" as expected because record levels of rainfall during construction severely delayed the project.

Conclusion and Discussion

To summarize, this article evaluates the effectiveness of nine PPP highway projects in Virginia that were developed and opened to traffic between 1990 and 2016. The analysis shows that Virginia has used highway PPP primarily for financing or risk reduction, and much less for efficiency gains. The examination of PPP contract agreements shows that not all PPP projects involved private finance, tolling, or concession. An example of public-finance PPP that is distinctive to the U.S.A. is DB, in which the private contractor assumes the combination of design and construction roles to reduce the risks of cost overrun and construction delay. For PPP projects that involved private finance, there has been a gradual shift from primary private finance to partnership finance. Later projects tended to partner with more professional developers, involved more private equity, and had longer concession terms. The allocation of project risks has also been adjusted over time. While construction-related risks were always allocated to private partners, the government retained administrative or regulatory risk in the public sector for more recent projects, and became more successful in shifting revenue risk to private developers in recent toll projects. Overall, the analysis shows that these PPP arrangements in Virginia have been largely effective in accessing innovative finance and preventing cost overrun, while there is limited evidence of success regarding shifting revenue risk or achieving efficiency gains.

This evaluation of the Virginian experiences also has practical implications for other states that are pondering whether and how to use PPP in highway development. First, private-finance PPP models, especially DBFOM, prove to be an effective way of attracting private finance. With proper arrangements, professional developers are willing to contribute some initial funding, either through private sector debts or private equity, in exchange for the concession rights to operate the facility for a long period of time to recover their investment. Second, relatively simple PPP arrangements that do not involve private finance, such as DB, can be used to significantly reduce the risks of cost overrun and, to a lesser extent, construction delay, by allocating these risks to the private contractor that directly manages facility construction. Third, care should be taken to manage revenue risk properly, which requires suitable provisions in the authorizing legislation and the selection of a competent private partner. Lastly, these Virginian PPP cases provide limited evidence about the intention of efficiency gains or the achievement of it, which is more emphasized in the UK's PFI, as evidenced by the use of VfM analysis.

There may exist inherent tradeoffs between multiple goals of PPP. Attracting private finance can reduce the short-term pressure on fiscal revenue or public borrowing, but it will incur higher financial cost, because public-sector borrowing (for example, through municipal bonds or SIB) can be much cheaper. Preventing cost overrun, construction delay, or revenue shortfall will enhance budgetary control, but these arrangements do not always come with lower private bids. As the Virginian experiences show, PPP highway projects could effectively attract private finance and prevent cost overrun, but they may be less successful in shifting revenue risk and achieving efficiency gains. How to manage the tradeoffs between these and other PPP goals would be an important question for future studies.

Acknowledgments

Funding for this research was provided by Projects of International Cooperation and Exchanges, National Natural Science Foundation of China (No. NSF71661137004).

Author Contributions

The authors confirm contribution to the paper as follows: study conception and design: Zhirong Jerry Zhao, Yin Wang; data collection: Yin Wang; analysis and interpretation of results: Yin Wang, Zhirong Jerry Zhao; draft manuscript preparation: Zhirong Jerry Zhao, Yin Wang. Both authors reviewed the results and approved the final version of the manuscript.

References

1. Zhao, Z. J. *Advancing Public Interest in Public-Private Partnership of State Highway Development*. Humphrey Institute of Public Affairs Report no. MN/Rc2011-09. Minnesota Department of Transportation, 2011.
2. FHWA. *P3 Home -P3 Projects*. Federal Highway Administration, Washington, D.C. https://www.fhwa.dot.gov/ipd/p3/p3_projects/. Accessed July 13, 2017.
3. Geddes, R. R., and B. L. Wagner. Why Do US States Adopt Public–Private Partnership Enabling Legislation? *Journal of Urban Economics*, Vol. 78, 2013, pp. 30–41.
4. Hodge, G. A., and C. Greve. Public-Private Partnership in Developing and Governing Mega-Projects. In *International Handbook on Mega Projects* (Priemus, H., and B. Van Wee, eds.), Edward Elgar, Cheltenham, UK, 2013, pp. 182–208.
5. Hodge, G. A., and C. Greve. Public–Private Partnerships: An International Performance Review. *Public Administration Review*, Vol. 67, No. 3, 2007, pp. 545–558.
6. FHWA. *Report to Congress on Public-Private Partnerships*, Federal Highway Adminstration, Washington, D.C., 2004.
7. Hodge, G. A. Reviewing Public-Private Partnerships: Some Thoughts on Evaluation. In *International Handbook on Public-Private Partnerships* (G.A. Hodge, C. Greve, and A.E. Boardman, eds.), Edward Elgar, Northampton, Mass., 2010, pp. 81–112.
8. Shaoul, J. The Private Finance Initiative or the Public Funding of Private Profit? In *The Challenge of Public-Private Partnerships: Learning from International Experience* (Hodge, G., and C. Greve, eds.), Edward Elgar, Northampton, Mass., 2005, p. 190.
9. Siemiatycki, M., and N. Farooqi. Value for Money and Risk in Public–Private Partnerships. *Journal of the American Planning Association*, Vol. 78, No. 3, 2012, pp. 286–299.
10. Dahler-Larsen, P. Evaluation and Public Management. In *The Oxford Handbook of Public Management* (Ferlie, E., L. E. Lynn, Jr., and C. Pollitt, eds.), Oxford University Press, 2005, pp. 615–639.
11. Van Den Hurk, M., and K. Verhoest. The Challenge of Using Standard Contracts in Public–Private Partnerships. *Public Management Review*, Vol. 18, No. 2, 2016, pp. 278–299.
12. Vining, A. R., A. E. Boardman, and F. Poschmann. Public–Private Partnerships in the US and Canada: "There are No Free Lunches." *Journal of Comparative Policy Analysis: Research and Practice*, Vol. 7, No. 3, 2005, pp. 199–220.
13. Flyvbjerg, B., M. S. Holm, and S. Buhl. Underestimating Costs in Public Works Projects: Error or Lie? *Journal of the American Planning Association*, Vol. 68, No. 3, 2002, pp. 279–295.
14. Das, T. K., and B. S. Teng. Trust, Control, and Risk in Strategic Alliances: An Integrated Framework. *Organization Studies*, Vol. 22, No. 2, 2001, pp. 251–283.
15. Wang, Y., and Z. Zhao. Motivations, Obstacles, and Resources Determinants of Public-Private Partnership in State Toll Road Financing. *Public Performance & Management Review*, Vol. 37, No. 4, 2014, pp. 679–704.
16. Gifford, J., and M. Transue. The Evolution of Virginia's Public-Private Partnership Enabling Statutes. *Journal of Corporation Law*, Vol. 41, No. 1, 2015, pp. 265–282.
17. Yin, R. K. *Case Study Research: Design and Methods*. Sage Publications, Thousand Oaks, Calif., 2013.
18. George, A. L., and A. Bennett. *Case Studies and Theory Development in the Social Sciences*. MIT Press, Cambridge, Mass., 2005.
19. FHWA. *Project Finance Primer 2010*. Federal Highway Administration, Washington, D.C., 2010.
20. FHWA. *Project Profile: Pocahontas Parkway*. Federal Highway Administration, Washington, D.C., https://www.fhwa.dot.gov/ipd/project_profiles/va_pocahontas.aspx. Accessed July 13, 2017.
21. Ybarra, S. J. *Private Sector Participation in Transportation*. Testimony before House Committee on Goverment Reform Subcommittee on Energy Policy, Natural Resources and Regulatory Affairs, September 30. 2004.
22. Regimbal, J. J. *An Analysis of the Evolution of the Public-Private Transportation Act of 1995*. Southern Environmental Law Center, Charlottesville, Va., 2004.
23. Boisseau, P., and S. Barry. Pocahontas Parkway Association Issues $350 Million in Bonds to Build Rt. 895 Connector as Toll Road. In *Business Wire*, 1998, www.businesswire.com. https://www.thefreelibrary.com/Pocahontas+Parkway+Association+Issues+%24350+Million+in+Bonds+to+Build...-a050156102. Accessed July 22, 2017.
24. Kozel, S. M. *Route VA-288 Construction - Western Section*. http://www.roadstothefuture.com/VA288_Construction_Western.html#288-PPTA. Accessed November 1, 2017.
25. Martz, M. Pocahontas Parkway to Change Control—Official Says Customers Using Pocahontas 895 Won't be Affected. In *Richmond Times-Dispatch*, BH Media Group, Inc., 2013.

The Standing Committee on Revenue and Finance (ABE10) peer-reviewed this paper (18-01316).

Article

TRR
JOURNAL OF THE TRANSPORTATION RESEARCH BOARD

Transportation Research Record
2018, Vol. 2672(4) 54–64
© National Academy of Sciences:
Transportation Research Board 2018
Article reuse guidelines:
sagepub.com/journals-permissions
DOI: 10.1177/0361198118790125
journals.sagepub.com/home/trr
SAGE

The Tradeoffs of Tolling Untolled Roads

Ronald Davis[1], Yogesh Patel[2], Christopher Mwalwanda[3], and Edward Regan[4]

Abstract

Recent years have seen increasing study and implementation of imposing tolls on existing untolled roadway capacity. This is due to growing agency budgetary challenges, increasing infrastructure needs, and some easing of legislative restrictions and pilot-project opportunities to add tolls to untolled roads. This paper presents a summary of current policy, profiles of facilities in operation, and a discussion of common tradeoffs related to tolling untolled roads. The facilities included are the Port Mann Bridge, Snapper Creek Expressway, Ohio River Bridges, Dominion Boulevard, Downtown Tunnel, Midtown Tunnel, SR 520 Bridge, and Tacoma Narrows Bridge. The tradeoffs discussed include those related to traffic diversion, the argument that untolled roads have already been paid for, that new toll payers may be unfamiliar with how to pay, equity, how tolling provides acceleration of project implementation and a dedicated funding source, and all-electronic tolling. Considering traffic diversion, the facilities included in the study were found to have from 10% to 36% less traffic after tolling, with most in the range of 15% to 25% less. Considering project acceleration, tolling on all facilities was established to fund major improvement projects that likely would not have been possible without toll revenue. Given consideration of all tradeoffs, it is important to take a long-term approach to toll project evaluation.

Transportation agencies are increasingly turning to many different sources to address their funding needs. There has also been increasing use of various financing mechanisms in contrast to more traditional pay-as-you-go funding (*1*). Regarding funding and financing through tolls, there has been a long history in the United States of financing new greenfield roadway projects with future toll revenue. For example, this was especially the case leading up to the passage of the 1956 Federal-Aid Highway Act (*2*).

Recent decades have also seen a resurgence of toll financing of new roadway projects, especially in Florida and Texas, to address the transportation needs of growing populations beyond the initial construction of the Interstate Highway system (*3*). When using toll financing, tolls have commonly continued beyond the initial financing horizon (typically 30 years). This provides funding and financing for ongoing maintenance, rehabilitation, reconstruction, and expansion of the facility itself as well as the construction of other new toll facilities operated by the same toll agency.

Aside from the traditional use of toll revenue for new toll facilities, there has been some easing of legislative restrictions and pilot-project opportunities in the past few federal transportation bills which facilitate imposing tolls on untolled roads. Recent years have also seen growing budgetary challenges for many transportation agencies. Decrease in fuel tax revenues, due to improving fuel efficiency of vehicles and many examples of fuel tax rates not keeping pace with inflation, is occurring at the same time as needs are growing due to aging systems. These factors have led to increasing study and implementation of tolls on existing untolled roadway capacity to fund and finance maintenance, rehabilitation, reconstruction, and additional capacity (*4*, *5*).

There now are several examples of projects involving adding tolls to previously untolled roadway and bridge capacity in operation in the United States and Canada. However, advancing toll projects despite a lack of alternative funding options continues to be cited as a challenge in transportation funding and finance (*1*). To provide more background information to this discussion, this paper presents a review of implemented tolling untolled road projects. A summary of current policy and common tradeoffs when tolling untolled roads is also

[1]CDM Smith, Lansing, MI
[2]CDM Smith, New Haven, CT
[3]CDM Smith, Austin, TX
[4]CDM Smith, Maitland, FL

Corresponding Author:
Address correspondence to Ronald Davis: davisrw@cdmsmith.com

included. This paper is intended to aid transportation agencies, decision makers, and researchers who are considering studies to add tolls to toll-free capacity.

There have been many examples of price-managed lane projects where tolls were added to existing, high occupancy vehicle (HOV) lanes to form high occupancy and toll (HOT) lanes, or where new, priced lanes were added to existing highways. These projects are not the focus of this paper as their tradeoffs and lessons learned are somewhat different to the process of adding tolls to existing full roadways and bridges. Other literature is available to discuss the tradeoffs of price-managed lanes (*6*).

Current United States Federal Tolling Policy

Generally, current federal law prohibits the collection of tolls on Interstate Highways; however, in recent years, exceptions to this tolling prohibition have been introduced (*7*). Federal law is less restrictive about imposing tolls on non-Interstate Highways. Beginning with the enactment of ISTEA in 1991 and through succeeding federal surface transportation authorization acts since then, several exceptions have been provided under Section 129 (23 USC Section 129 – general tolling provisions) and through special pilot programs, such as the Interstate System Reconstruction and Rehabilitation Pilot Program (ISRRPP) and Value Pricing Pilot Program (VPPP). Under the VPPP and ISRRPP, a state must execute a cooperative toll agreement with FHWA to be authorized to impose tolls on currently untolled facilities.

Under current federal law, new highways, bridges, and tunnels (including on the Interstate System) can be constructed as tolled facilities; new tolled lanes can be added to existing highways (as long as the number of existing toll-free lanes is not reduced); bridges and tunnels can be reconstructed or replaced as tolled facilities; and capital improvements can be made to existing tolled facilities with federal funds. States are also allowed to institute tolls on any road that is not an Interstate Highway if it is being reconstructed or if it is a new construction project. If the non-interstate road is not part of a new construction or reconstruction project, tolling is allowed only if it is part of VPPP.

ISRRPP allows up to three existing and currently free Interstate Highways to be converted to tolled facilities in order to fund needed reconstruction or rehabilitation of the facilities that would not otherwise be possible without toll revenues. Several states have been chosen to participate in this pilot program by FHWA, and have performed significant tolling-related studies. However, tolling has not yet been implemented as part of ISRRPP.

VPPP was initially authorized in ISTEA (as the Congestion Pricing Pilot Program) and continued in subsequent surface transportation authorization acts. It encourages implementation of a variety of strategies to manage highway congestion, including tolling. Under VPPP, tolls may be imposed on existing toll-free highways, bridges, and tunnels, so long as variable pricing is used to manage demand for the facility.

Project Examples

There are several examples of projects involving adding tolls to untolled roadway and bridge capacity in operation in the United States and Canada with others being close to implementation. This section gives some background information on several of these projects, which are shown in Figure 1.

Implemented Projects

Nine facilities were identified in the United States and Canada for this study as having implemented tolling on untolled roads, bridges, or tunnels. These facilities are listed in Table 1 with the impacts of tolling in terms of average daily traffic. The date tolling started, the number of lanes before and after tolling, and 2016 traffic are also included in Table 1. Considering the impact on traffic, the facilities were found to have in the range of 10% to 36% less traffic, with most in the range of 15% to 25% less. As shown, even with substantial capacity increases on some facilities, tolling still caused traffic diversion. It should also be noted that, while the overall impacts of tolling are shown in Table 1, the rate of traffic diversion can vary substantially on certain days and time periods. For example, weekday average traffic on the SR 520 Bridge is now higher during both directional peak periods (westbound evening peak and eastbound morning peak) compared with traffic before tolling. However, when also considering weekday off-peak and weekend traffic, overall average traffic is still about 25% lower. Data that would permit similar time of day comparisons with other facilities was not readily available for this study, but other facilities most likely would show less diversion in peak periods compared with off-peak.

Impact on traffic at different facilities can vary for a variety of reasons, including: the relative expense of the toll rates implemented, the level of toll rates drivers in the region are willing to pay, the characteristics of improvement projects funded by the toll revenue, the amount of congestion on the facility, and the availability of alternative travel options. Another important consideration is the impact of tolling on construction. For most of these facilities, tolling helped to fund major improvement projects, which normally had a significant impact on traffic. An effort was made to include data from years before and after the main construction impacts in Table 1.

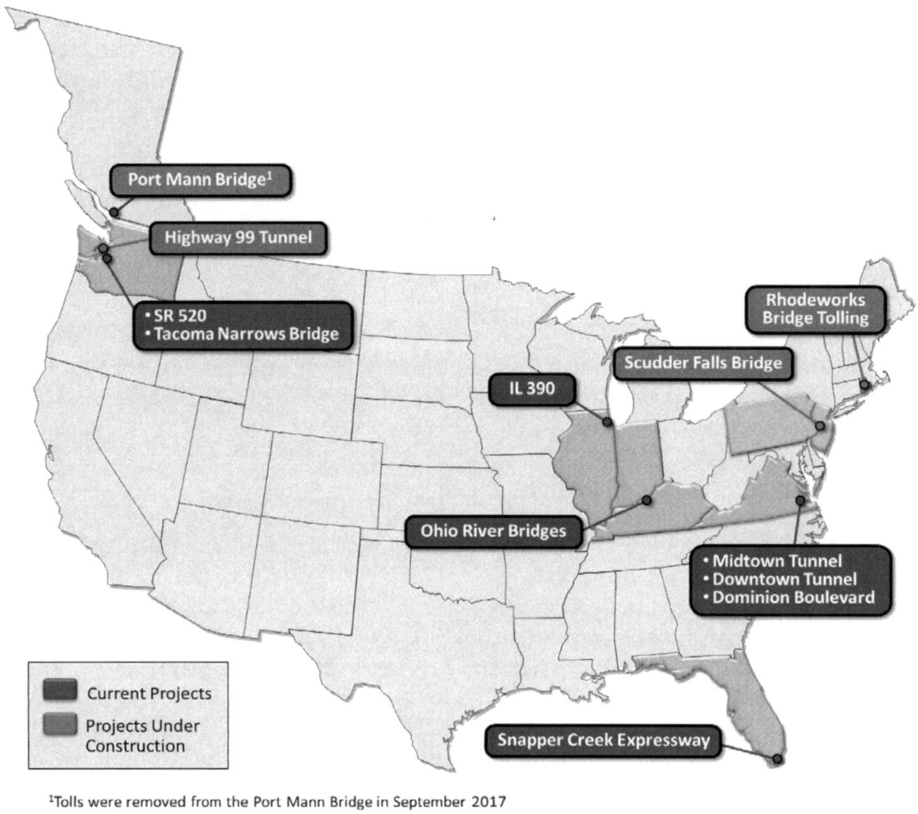

Figure 1. Map of projects involving tolling previously untolled roads, bridges, and tunnels as of January 1, 2018.

However, this causes some of the comparisons to be made with data several years apart which can, in turn, cause the comparisons to be somewhat influenced by changes in travel demand over time in the corridors. No attempt was made to adjust the comparisons for underlying travel demand changes because of challenges in identifying the correct travel demand adjustment for different facilities, regions, and time periods. A more detailed study that also accounts for changes in travel demand could result in some small variation in the findings on traffic diversion at individual facilities compared with those presented here.

Table 2 shows impact on travel time for the facilities, using before and after tolling speed data available from INRIX. Average weekday travel times were calculated for representative roadway segments with segment distances included in the table. Considering impact on travel time, all four facilities with data available were found to have improved speeds in the morning peak (in the peak direction), with average travel time improvements ranging from 2.6 to 5.4 minutes for the segments analyzed. In the evening peak three facilities had improved travel times, ranging from 2.1 to 4.3 minutes. Travel time improvements are due to both lower traffic after tolling and the capacity improvements made to the different facilities. One facility (SR 520 Bridge) was found to have increased travel times, calculated to be 5.1 minutes. This outlier is discussed in more detail below. Impact on travel time can vary across different facilities for similar reasons as discussed for impact on traffic volume in the previous paragraph.

Project profiles, including background details, financial information, and other important considerations for all facilities in Table 1, are provided in the next subsections.

British Columbia: Port Mann Bridge. The Port Mann Bridge is part of Highway 1 in the British Columbia section of the Trans-Canada Highway, linking Vancouver with the rest of Canada. A new bridge, with four general purpose lanes and one HOV lane in each direction, replaced the previous five-lane bridge. The previous bridge, constructed in 1964, had two general purpose lanes in each

Table 1. Diversion Impacts of Implemented Examples of Tolling Untolled Roads

State/ Prov	Facility	Tolling started	Before tolling Year	Before tolling Lanes	Before tolling Traffic	% change Traffic	After tolling Year	After tolling Lanes	After tolling Traffic	2016 traffic
BC	Port Mann Bridge[a]	12/8/2012	2008	5	118,602	−17%	2013	10	98,819	111,967
FL	Snapper Creek Expressway	7/17/2010	2009	4	43,748	−20%	2011	4	34,790	37,130
IL	IL 390 (US 20 to I-290)[b]	7/5/2016	2012	4	191,450	−23%	Jul-Oct 2016	4–6	146,550	–
IN-KY	Downtown Crossing	12/30/2016	2012	7	134,865	–	Apr–Jun 2017	12	68,902	–
	East End Crossing		–	–	–	–		4	17,423	–
	Total Ohio River Bridges[c]		2012	7	134,865	−36%		16	86,325	–
VA	Dominion Boulevard Veterans Bridge	2/9/2017	2012	2	29,000	−24%	Mar 2017	4	21,907	–
VA	Downtown Tunnel	2/1/2014	2012	4	91,000	−16%	2016	4	76,000	76,000
VA	Midtown Tunnel	2/1/2014	2012	2	35,000	−20%	2016	4	28,000	28,000
WA	SR-520 Bridge	12/29/2011	2011	4	93,000	−25%	2016	6	70,000	70,000
WA	Tacoma Narrows Bridge	7/16/2007	2002	4	86,000	−10%	2008	7	77,000	83,000

[a]Tolls were removed on September 1, 2017 causing an increase in traffic of about 27% compared with September 2016.
[b]Traffic for IL 390 is the total sum of average daily traffic at three toll gantry locations between US 20 and I-290. After tolling this would be the same as total average daily transactions.
[c]Refer to the "Kentucky and Indiana – Ohio River Bridges" section for more information on this comparison.
Sources: (*8–15*).

Table 2. Impact on Travel Time (TT) for Implemented Examples of Tolling Untolled Roads

State	Facility	Tolling started	Segment distance (mi)	Before tolling Year	Before tolling Average TT (min)	Change (min)	After tolling Year	After tolling Average TT (min)
Morning peak (07.00–09.00) Weekday average in peak direction								
IL	IL 390 (US 20 to I-290)	7/5/2016	6	2012	9.2	−3.5	Oct 2016	5.8
IN-KY	Downtown Crossing	12/30/2016	12	2012	17.7	−5.4	Apr–Jun 2017	12.3
VA	Downtown Tunnel	2/1/2014	6	2012	16.0	−3.2	2016	12.8
WA	SR-520 Bridge	12/29/2011	7	2011	11.6	−2.6	2016	9.0
Evening peak (16.00–18.00) Weekday average in peak direction								
IL	IL 390 (US 20 to I-290)	7/5/2016	6	2012	8.1	−2.1	Oct 2016	6.0
IN-KY	Downtown Crossing	12/30/2016	12	2012	16.4	−4.0	Apr–Jun 2017	12.3
VA	Downtown Tunnel	2/1/2014	6	2012	15.1	−4.3	2016	10.8
WA	SR-520 Bridge	12/29/2011	7	2011	14.5	5.1	2016	19.6

Source: (*16*).

direction and an eastbound HOV lane that was added in 2001. The new bridge was constructed as part of the overall 37-kilometer Port Mann/Highway 1 Improvement Project. This project also included widening and interchange upgrades to Highway 1. Severe congestion in the corridor was the main driver of the improvement project.

In 2008, the Transportation Investment Corporation (TIC), a Crown corporation (Canadian state-owned enterprise), was formed to deliver the Port Mann/Highway 1 Improvement Project. After receiving proposals for public-private partnerships (P3s) for the project in 2008, TIC entered a P3 agreement-in-principle in

January 2009 with the Connect BC Development Group. However, they were unable to reach a final agreement and the project was changed to a design–build with TIC (via Province of British Columbia debt) providing the project financing. The full cost of the Port Mann/ Highway 1 Improvement Project was $3.3 billion (Canadian dollars) which was planned to be repaid by toll revenue from the Port Mann Bridge over a 40-year debt term (*17*). The new bridge first opened to traffic in September 2012 and began tolling in December 2012. Tolls were removed in September 2017 (along with tolls on another bridge in the region) in fulfillment of a campaign promise made in the 2017 British Columbia general elections. The reasons given for removal of toll were related to equity and fairness, as some parts of the region had better access to other existing untolled crossings than those living closer to the Port Mann Bridge. With the removal of tolls, the outstanding bridge debt was transferred to general provincial debt.

Traffic on the Port Mann Bridge declined by about 17% (about 19,800 vehicles) with tolling when comparing pre-construction traffic from 2008 with traffic in 2013. Traffic after 2013 grew at a steady pace to nearly 112,000 vehicles per day in 2016. After the removal of tolls, there were 145,000 vehicles per day in September 2017, compared with 114,000 in September 2016—an increase of about 27%, or 31,000 vehicles. The significant increase in traffic has caused new congestion in the corridor.

Florida: Snapper Creek Expressway. The Snapper Creek Expressway is a 2.5-mile section of the Miami-Dade Expressway Authority (MDX) toll road system that consists of four other toll facilities in southeastern Florida. The expressway was originally opened untolled in 1980 by the Florida Department of Transportation (FDOT) at a cost of $8.3 million. The roadway was planned to be opened by FDOT along with the Homestead Extension of the Florida's Turnpike, the Dolphin Expressway (SR 836) Extension, and Don Shula Expressway (SR 874) in the mid 1970s but was delayed as funding from the bonds sold for construction of the four roads ran short (*18*). In the mid 1990s FDOT transferred control of the Snapper Creek and four other toll facilities to MDX.

Before 2010, many trips on MDX system toll roads, including trips on the Snapper Creek Expressway, could be made without paying a toll. Between 2010 and 2014 MDX implemented a new all-electronic tolling (AET) system which included many new tolling points, including one on the Snapper Creek Expressway in 2010. Revenue from the additional tolling locations was used to fund several major construction projects on the MDX system.

Illinois: IL 390. IL 390 was first opened in 1993 by the Illinois Department of Transportation (IDOT) as the 6.5-mile Elgin O'Hare Expressway in the western Chicago suburbs. In 2012, an agreement between IDOT and the Illinois Tollway was made to convert the facility to toll road managed by Illinois Tollway which would make several improvements and extensions. The project has included widening and rehabilitation to the existing Elgin O'Hare Expressway (now called IL 390), extension of the expressway to I-290 by grade separating existing interchanges, and improvements to the I-290 interchange. These elements opened in spring 2016 and tolling began on IL 390 between US 20 and I-290 on July 5, 2016. An extension of the IL 390 toll road east of I-290 also began tolling on November 1, 2017. Construction of north–south segments of the project (around western O'Hare Airport) are also planned to begin in the coming years and be fully opened by 2026.

The overall project, called the Elgin O'Hare Western Access (EOWA), is estimated to cost $3.4 billion. Of this total, $3.1 billion is being funded by the Illinois Tollway and $ 0.3 billion is anticipated from the local DuPage County (*19*). The Illinois Tollway portion is funded by current toll revenue and toll revenue bonds which are backed by future toll revenue. The toll revenue for the EOWA and other systemwide Illinois Tollway projects comes from the entire Illinois Tollway system which includes four other toll road facilities in addition to IL 390.

The impact on traffic between 2012 (the year before construction started) and July to October 2016, following the opening and tolling of the initial IL 390 toll road (from US 20 to I-290), was a decrease of 23%. This figure is based on total combined traffic at the three toll plaza locations. In addition to increases from normal future demand growth, traffic will likely increase on IL 390 between US 20 and I-290 in the future as additional phases of the EOWA project are opened. Travel time improvements have been observed on IL 390 due to lower traffic and the corridor widening and access control improvements.

Kentucky and Indiana: Ohio River Bridges. The Ohio River Bridges project between Louisville, Kentucky, and Indiana consisted of two major project elements: the $1.5 billion Downtown Crossing project and the $1.3 billion East End Crossing project. The purpose of the overall project was to increase cross-river capacity, improve safety, and complete the eastern suburban loop around Louisville. The East End Crossing project, which was led by Indiana, included new approach roads and a new Lewis and Clark Bridge over the Ohio River. The Downtown Crossing project, which was led by Kentucky, consisted of a new northbound-only I-65 Bridge built adjacent to the existing and previously untolled I-65 Kennedy Bridge, conversion of the existing

I-65 Kennedy Bridge to southbound-only travel, and improved access interchanges just north and south of the bridges. Tolling began on all three bridges on December 30, 2016.

The Downtown Crossing portion funding included Kentucky state and federal funding ($342 million), Kentucky GARVEE Bonds ($337 million), Indiana state and federal funding ($36 million), Kentucky toll revenue bonds ($272 million), and Kentucky TIFIA and Bond Anticipation Notes ($494 million) (*20*). Thus, about half of the Downtown Crossing project funding (toll revenue bonds and TIFIA) was backed by toll revenues.

As shown in Table 1, average daily traffic from 2012, which was the year before construction started, compared with early traffic results after tolling, showed a 36% decrease in traffic. This includes traffic on all new bridges as it is likely a sizeable amount of traffic formerly on I-65 is now using the new Lewis and Clark Bridge to cross the Ohio River. This impact may become less negative as drivers become more accustomed to paying tolls and some secondary construction projects are completed. Significant improvements in travel time have been observed on the Downtown Crossing due to lower traffic on I-65, the additional bridge capacity, and improvements to the bridge approaches.

Virginia: Dominion Boulevard Veterans Bridge. This project, owned by the City of Chesapeake, involved improving about four miles of the Dominion Boulevard over and approaching the southern branch of the Elizabeth River. To improve traffic operations, an aging two-lane steel lift bridge was replaced with the four-lane expressway standard Veterans Bridge with sufficient vertical clearance. The approach roads on both sides of the bridge were also widened and improved. Construction began in 2013, was substantially completed in November 2016, and tolling began February 9, 2017.

Funding for the $399 million project included a Virginia Transportation Infrastructure Bank (VTIB) loan ($152 million), toll revenue bonds ($159 million), state and federal funds ($86 million), and existing reserves and earnings ($3 million). Both the VTIB and toll revenue bonds are backed by system toll revenue from both the City of Chesapeake's toll facilities: the Dominion Boulevard and the Chesapeake Expressway (*21*).

As shown in Table 1, average daily traffic from 2012 on the Veterans Bridge, which was the year before construction started, compared with early traffic results after tolling showed a 24% decrease in traffic. This impact may become less negative as drivers become more accustomed to paying tolls and higher traffic during the summer months are included in the after-tolling data.

Virginia: Downtown and Midtown Tunnels. Tolls were added to the untolled Downtown (I-264) and Midtown (US 58) Tunnels in the Hampton Roads area of Virginia in February 2014 to help fund a second Midtown Tunnel tube, rehabilitation of the existing Midtown Tunnel, rehabilitation of the existing Downton Tunnel, and an extension of the Martin Luther King (MLK) Freeway. Financial close between the Virginia Department of Transportation (VDOT) and the private company Elizabeth River Crossings (ERC) occurred in April 2012 for a concession length of 58 years. The tunnels were originally tolled from their initial opening to the late 1980s when tolls were removed.

Construction activities started in early 2013 with construction on the new west Midtown Tunnel. Both lanes of the new tunnel first opened in August 2016. Rehabilitation of the existing Midtown Tunnel began in June 2016 and was completed June 30, 2017. Rehabilitation on the west Downtown Tunnel began with full weekend closures starting in August 2013 and on the east Downtown Tunnel with full weekend and weeknight single lane closures starting in July 2014. Substantial completion of the Downtown Tunnel Rehabilitation was achieved in August 2016. After construction began in May 2015, the MLK extension project opened in November 2016.

The total project cost was $2.1 billion which was planned to be funded from a $422 million TIFIA loan, $675 million in Private Activity Bonds, $272 million in equity contributions from ERC, $268 in toll revenues during construction, $43 million in TIFIA capitalized interest, and the remaining from VDOT contributions. Before financial close, VDOT's contribution was estimated to be $362 million but lower interest rates reduced the contribution needed to about $309 million. An additional contribution was subsequently made by VDOT of about $113 million to keep toll rates significantly lower over the life of the project than originally estimated. In 2014, VDOT contributed an additional $83 million to the project to keep toll rates lower during the construction period (*22*). In July 2015, VDOT contributed an additional $78 million to the project to buy out the tolls on the MLK extension. Tolling was originally planned to be added to the MLK extension upon its completion (*23*). Public and political sentiment to minimize the tolling impact has been reflected in the additional investments by VDOT in the project to limit tolls.

As shown previously, the impact on traffic in the tunnels due to tolling was estimated to be a decrease of 15–20%. However, ongoing construction in 2016 when the most recent data was available may have been continuing to depress traffic volume. Another study that analyzed the impacts on traffic from tolling found similar impacts on the Downtown Tunnel but lower impacts on the Midtown Tunnel compared with this study (*24*). Considering available data on travel time, the Downtown Tunnel was calculated to have a 3.2-minute improvement in travel time in the morning peak and a 4.3-minute improvement in the evening peak.

Washington State: SR 520. The SR 520 Bridge is one of two floating bridges, along with the I-90 floating bridge, over Lake Washington east of Seattle. The original bridge opened in 1963 and was financed by tolls which were removed in 1979 after the original bonds were paid off. Planning for a new bridge and other corridor enhancement began in the 1990s to address the aging floating bridge structure and growing mobility needs in the region. Tolling was re-introduced on the original SR 520 bridge in December 2011 to help pay for the overall SR 520 Bridge Replacement and HOV Program, which began in 2012. In 2014 improvements on SR 520 east of the bridge were completed. The new SR 520 floating bridge opened on April 25, 2016. Construction was due to continue on the western approach bridge to the floating bridge through 2017 and on SR 520 to I-5 afterward.

Funding for the $4.56 billion overall project came from many sources. About 26% of the funding was from tolling-related sources. These are toll bonds ($660 million), a TIFIA loan ($300 million), deferred construction sales taxes to be repaid by toll revenues ($160 million), net toll revenue during construction ($70 million), and a portion of civil penalties collected from toll violations ($14 million). Non-tolling-related sources are GARVEE bonds ($925 million), state funding from several programs ($2,232 million), and other, primarily federal, funding ($300 million) (*25*).

Traffic on the SR 520 Bridge declined from 93,000 in 2011 (the year before construction and tolling began) to 70,000 in 2016 (the year the new bridge opened), a decrease of about 25% or 23,000 vehicles. Traffic continues to be affected by ongoing construction projects adjacent to the bridge. Much of the traffic diverted from SR 520 has been found to be using the parallel I-90 bridge. Traffic increased from 118,000 to 131,000 between 2011 and 2016 on the I-90 bridge, or by about 13,000 vehicles. Analysis of travel times on SR 520 showed improvements in the morning peak but longer travel times in the evening peak between 2011 and 2016. It is likely that the worse conditions in the evening peak are due to ongoing SR 520 construction and significant growth in travel demand in the region in recent years. The growth in travel demand has increased congestion on SR 520 east of Lake Washington and caused queuing back to SR 520 from congestion on I-5 west of Lake Washington.

Washington State: Tacoma Narrows Bridge. The second Tacoma Narrows Bridge opened in October 1950, replacing the original bridge which collapsed in November 1940. Both bridges were tolled, with the second having tolls removed in 1965 following the bill being passed to retire the bonds. To address growing congestion and safety concerns on the bridge, construction of a second span began in October 2002, which opened and began tolling in July 2007. The overall project also included improvements to the approaches on SR 16 and upgrades to the existing bridge.

The project had a final cost of $735 million (originally estimated at $849 million). Construction followed many years of study and controversy on the project which at one time included plans for private operation of the bridge in a design–build–finance–operate P3 model. In 2002, this was changed to a design–build model with Washington State Department of Transportation (WSDOT) assuming financing and operations. Funding came mostly from sales of bonds (motor vehicle fuel tax general obligation bonds) and from a small loan from the Washington State Motor Vehicle Account, both to be repaid by toll revenues (*26*). Related to toll rates, a citizen advisory commission was created to advise WSDOT on toll rate policy.

Traffic was estimated to decline about 10% due to tolls when comparing pre-construction levels in 2002 with after construction levels in 2008. The low diversion rate compared with other examples of tolling untolled roads is partially due to the absence of nearby alternative routes.

Projects under Construction or Advanced Development

Three examples of tolling untolled capacity-related projects are nearing implementation in the coming years: the Scudder Falls Bridge which crosses the Delaware River between Pennsylvania and New Jersey, the Rhodeworks bridge reconstruction program in Rhode Island, and the Highway 99 Tunnel in Seattle.

The Scudder Falls Bridge is one of 20 bridges over the Delaware River operated by the Delaware River Joint Toll Bridge Authority. Currently, tolls on seven vehicular bridges support operations and maintenance costs on all 20 authority bridges. Construction began in 2017 to replace the current four-lane, untolled Scudder Falls Bridge with a six-lane, tolled facility. Tolls are necessary due to the high cost of the project ($328 million) relative to the agency's current toll revenue and costs. Tolls are anticipated to be added in 2019.

Rhodeworks is a capital program to repair Rhode Island's deteriorating bridges and was signed into law in early 2016. The program will borrow against future federal aid money and against toll revenue. The toll revenue will be collected by adding tolls for heavy commercial vehicles on 14 existing untolled bridges around the state.

The funding will support investment into more than 650 bridges around the state. The program has generated considerable opposition from commercial vehicle organizations. The first two toll gantries were installed in February 2018 and the equipment is being tested as of March 2018.

The $3.2 billion Highway 99 Tunnel in downtown Seattle is a tunnel replacement of an above-ground elevated roadway called the Alaskan Way Viaduct. Tolls are planned for the new tunnel to cover a small portion of the construction costs as well as ongoing operations and maintenance costs for the facility. The project has faced many construction-related delays but is expected to open and begin tolling in fall 2018.

Projects Studied

Some notable examples of ongoing studies of tolling untolled roads are listed below:

- Alabama: Tolls are being considered to help fund a proposed wider and higher replacement of the I-10 bridge in Mobile.
- Connecticut: With several mega reconstruction projects on the horizon and a shortfall of traditional funding sources, Connecticut is considering imposing tolls on several highways. Tolling would vary by congestion level as it would be added through the state's federal VPPP slot.
- Florida: Tolls are being considered on the Treasure Island Causeway in St. Petersburg to help fund bridge maintenance activities. Tolls were previously imposed but were removed in 2006.
- Indiana: Transportation funding legislation passed in April 2017 included a requirement to study the feasibility of tolling Interstate Highways and to seek a FHWA waiver to do so. The study was completed in 2017 and a larger, follow-up study is being planned for 2018.

Discussion of Tradeoffs

Several commonly identified tradeoffs of tolling untolled roads are described below.

Traffic Diversion

Due to the nature of pricing itself, a certain number of drivers will choose to take alternative routes or modes or choose not to make trips in response to a new toll. This occurs both when tolls are imposed on untolled roads and when toll rates are raised on existing tolled roads. Communities with alternative roads that receive the diverted traffic from the toll facility may become concerned about increasing congestion and large commercial vehicle traffic on their roads. Additionally, travelers diverting will experience increased travel times. The rate of diversion can vary substantially on different days and time periods, with peak periods normally experiencing less diversion than off-peak. While traffic diversion is most commonly considered a disadvantage of tolling, in some situations diversion is part of the objective, such as in the case of congestion pricing. With congestion pricing, tolls are set to manage demand and limit congestion on the tolled facility by encouraging travel at less congested times or on other routes.

Untolled Roads Have Already Been Paid For

This tradeoff is often used by opponents of tolling when arguing against a project. The argument normally refers to the original construction cost of the facility, but most commonly does not consider other full life cycle costs and the eventual cost of reconstruction. Considering infrastructure cost more broadly, the cost of major improvements or reconstruction can be higher than initial construction costs of a facility even in real terms.

New Toll Payers May Be Unfamiliar with How to Pay

Especially when implemented in regions without existing toll facilities, tolling requires public outreach and ongoing customer service to inform drivers about elements of toll systems they may not be familiar with. Users of current toll facilities often still associate tolling with traditional stop-and-go toll booths with cash payment. However, new toll facilities in the United States and Canada have all utilized AET in recent years. AET systems collect tolls at highway speeds using overhead gantries with electronic sensors and video equipment. Public outreach commonly includes explanation of ways to open a toll account, obtain an electronic transponder, and pay toll bills for frequent users. For infrequent users, public outreach commonly includes information on the billing process. This tradeoff can be challenging due to the cost and effort required for public outreach and customer service as well as potential resistance from some drivers to the added steps required to pay to use a toll facility.

Equity Considerations

As described previously in relation to the Port Mann Bridge, equity considerations can be a tradeoff relating to tolling untolled roads. As was the case with the Port Mann Bridge, transportation equity debates are often foremost about geography, with elected officials looking for policies that benefit their constituents. Income equality is also an important consideration with tolling (*27*). Equity concerns are sometimes addressed on toll facilities by implementing various discount programs, such as discounts for frequent users or residents of certain regions or states.

Acceleration of Project Implementation and a Dedicated Funding Source

Tolling allows financing backed by toll revenues to pay for projects that could otherwise not be fully funded, thereby significantly accelerating their completion. As described in the examples in this paper, in several cases of tolling untolled roadways, toll financing was combined with other sources of project financing and funding. Tolling is normally cited as critical to timely implementation of the project. In addition to project acceleration, tolling untolled roadways to fund maintenance can potentially save money over time if ongoing deferred maintenance makes a project more expensive in the future. This has been cited as one reason for tolling in the Rhodeworks program discussed previously (*28*).

All-Electronic Tolling

As described previously, AET systems have been implemented on all new toll facilities in the United States and Canada in recent years, including those that have involved tolling untolled roads. AET comes with its own set of tradeoffs. On the one hand, AET eliminates additional vehicle emissions and congestion caused by vehicles slowing down at traditional toll booths. AET also requires less right of way for initial toll plaza construction, which can lead to significant capital cost savings. On the other hand, AET requires agencies to oversee complicated collection and billing processes. Collecting tolls from all vehicles, including those without a transponder, has been found to be challenging, especially if a vehicle is registered out of state or out of the country. Several of the agencies operating the facilities described previously, most notably the Downtown and Midtown Tunnels, made changes to their collection and enforcement procedures after opening as a result of challenges with toll collection for vehicles without a transponder (*29*).

Cost of Toll Collection Is Higher than Other Traditional Sources of Funding

The cost of collecting toll revenue can include the cost of toll collection equipment, back office software and staff, and customer service systems. Traditional sources of funding, such as fuel taxes and vehicle registration fees, also have administrative costs associated with them but studies have shown the cost of collecting them to be less than the cost of collecting tolls (*30*).

Tolling Is Not Impacted by Increasing Fuel Efficiency or Alternative Vehicles

The declining sustainability of fuel taxes as a source of government revenue is often cited as a reason to consider alternative funding sources including tolling. At the federal level, more fuel efficient vehicles and electric vehicles are being encouraged, however this has the effect of limiting the fuel tax revenue collected. Because tolling is based on usage of the road itself, it is not affected by improvements in vehicle fuel systems technology.

Fuel Taxes Will Also Be Paid to Drive on a Toll Road

This tradeoff is also often used by opponents of tolling. Even if a toll facility is entirely supported by tolls, a driver will still be paying fuel tax, which is used to support other roads, when traveling on the toll facility. There have been discussions of reimbursing drivers for the fuel tax they pay when traveling on a toll facility but so far no such program has been implemented. One factor to consider with this tradeoff is that some projects, such as the Ohio River Bridges and SR 520 Bridge discussed above, include sources of revenue that are supported by fuel taxes. Another factor is that the amount of fuel tax incurred by drivers, especially on toll bridges and short toll roads, would be very small.

Conclusion

As can be inferred from the varied examples of tolling untolled roads described in this paper, there is no "one size fits all" approach. Tolling has been implemented in different parts of the country on road projects with widely varying characteristics. Despite their different characteristics, some overall observations can be made from the examples and discussion of tradeoffs.

First, traffic diversion is one of the most significant impacts of tolling untolled roads. The nine facilities included in this study were found to have overall average impacts of tolling in the range of 10% to 36% less traffic, with most in the range of 15% to 25% less. All four facilities with available INRIX speed data were found to have travel time improvements in the morning peak and three of the four facilities were found to have travel time improvements in the evening peak. Travel time improvements were due to both lower traffic after tolling and improvement projects funded or partially funded by toll revenue.

Another observation is that, in many of the examples, revenue for the significant improvement projects to the facility being tolled and/or near the facility being tolled also comes from other non-toll sources or they are also supported by revenue from other existing toll roads. For example, funding for the Downtown Crossing project included general state and federal sources in addition to toll revenue backed bonds. Also, bonds for the IL 390 project were backed by toll revenues from the entire Illinois Tollway system. With the high cost of projects, combining different funding sources is used to cover project costs in many situations.

On eight of the nine example projects included (with the possible exception being the Snapper Creek Expressway), it is likely the improvement projects supported by toll revenue would have been delayed many years without the revenue generated from tolling. Given their high cost, these improvement projects would have put significant strain on already limited agency budgets without tolling. Thus, project acceleration is perhaps the most significant benefit of tolling untolled roads.

Given some of the tradeoffs discussed, for example, traffic diversions, how new toll payers may be unfamiliar with how to pay, and how the cost of toll collection can be relatively high, it is important to take a long-term view when evaluating tolling untolled road-type projects. The value of accelerating replacement of a critical piece of infrastructure by implementing tolling may not be realized in a short-term evaluation. Tolling also makes more sense when taking a long-term view, since improvements in vehicle fuel economy and electric vehicles are putting an increasing strain on fuel tax revenues, a challenge that is expected to become more severe into the future.

Author Contributions

The authors confirm contribution to the paper as follows: study conception and design: all authors; data collection: Davis, Patel; analysis and interpretation of results: Davis, Patel, Mwalwanda, Regan; draft manuscript preparation: Davis, Patel. All authors reviewed the results and approved the final version of the manuscript.

References

1. *Transportation Governance and Finance: A 50-State Review of State Legislatures and Departments of Transportation.* American Association of State Highway and Transportation Officials, Washington, D.C., 2016.
2. Owen, W., and C. Dearing. *Toll Roads and the Problem of Highway Modernization.* The Brookings Institution, Washington, D.C., 1951.
3. Bustillo, M., and N. Koppel. In Texas, Toll Roads Proliferate – and a Backlash Builds. *Wall Street Journal.* October 20, 2014.
4. *U.S. Interstate Tolling Merited but Warrants Caution.* Fitch Ratings, New York, 2017.
5. As Gas-Tax Profits Decline, More States May Turn to Tolls. *Governing.* January 30, 2017.
6. *NCHRP Report 835: Guidelines for Implementing Managed Lanes.* Transportation Research Board of the National Academies, Washington, D.C., 2016.
7. *Current Laws on Tolling Existing and New Facilities on Federal Interstate Highways.* Connecticut Department Transportation, 2015. https://www.cga.ct.gov/2017/tradata/od/2-24-17%20%20Current%20Laws%20on%20 Tolling%20 Existing%20and%20New%20Facilities%20on%20Federal %20Interstate%20Highways%20,%20CDM-Smith%20.pdf. Accessed July 1, 2017.
8. *Traffic Data Program: Traffic Data For: Port Mann – P-16-2EW – N.* British Columbia Ministry of Transportation. http://www.th.gov.bc.ca/trafficData/tradas/tradas.asp?loc=P-16-2EW. Accessed May 10, 2017.
9. *2016 Historical AADT Report: Site: 0193 – SR-878, 0.2 MI W SR 826, DADE CO.* Florida Department of Transportation Statistics Office. https://flto.dot.state.fl.us/website/FloridaTrafficOnline/viewer.html. Accessed May 10, 2017.
10. Are Tolls on Former Elgin-O'Hare Shifting Traffic Elsewhere? *Daily Herald.* March 13, 2017.
11. *Traffic Count Database System: Location ID 971010.* Indiana Department of Transportation. http://indot.ms2soft.-com/tcds/tsearch.asp?loc=Indot&mod=. Accessed May 2, 2017.
12. *Crossings Climb, Call Times Improve In Second Quarter.* RiverLink. https://riverlink.com/2017/07/18/crossings-climb-call-times-improve-in-second-quarter/. Accessed July 31, 2017.
13. *Average Daily Traffic Volumes with Vehicle Classification Data on Interstate, Arterial, and Primary Routes: 2016.* Virginia Department of Transportation Traffic Engineering Division, Richmond, VA, April 18, 2017.
14. *Chesapeake Transportation System Disclosure Reports.* City of Chesapeake Virginia. http://www.cityofchesapeake.net/government/city-departments/departments/Public-Works-Department/chesapeake-transportation-system/bridges-dominion-blvd-improvements/chesapeake-transportation-system-disclosure-reports.htm. Accessed May 11, 2017.
15. *Annual Average Daily Traffic Volumes in XLS Format.* Washington State Department of Transportation. https://www.wsdot.wa.gov/mapsdata/travel/docs/2000-2016-ATRs.xls. Accessed May 10, 2017.
16. *Speed Using IRIX Data.* INRIX. https://inrix.ritis.org/analytics/. Accessed July 14, 2017.
17. *Premier Launches Construction of New Port Mann Bridge.* News Release. Ministry of Transportation and Infrastructure, British Columbia. February 4, 2009.
18. Snapper Creek: U.S. 1-South Dade Connection Finally Open. *Miami News.* March 19, 1980.
19. *Elgin O'Hare Western Access Project.* Illinois Tollway. https://www.illinoistollway.com/documents/20184/106941/2016 + EOWA + Overview_Factsheet/57f5cbed-4c1c-4954-92d6-b670ecf1931f. Accessed June 5, 2017.
20. *Louisville – Southern Indiana Ohio River Bridges Project Financial Plan – 2016 Annual Update.* Kentucky

Transportation Cabinet and Indiana Department of Transportation, 2016.
21. *Project Facts*. City of Chesapeake Virginia. http://www.cityofchesapeake.net/government/City-Departments/Departments/Public-Works-Department/chesapeake-transportation-system/bridges-dominion-blvd-improvements/project-facts.htm. Accessed May 23, 2017.
22. *Downtown Tunnel/Midtown Tunnel/MLK Extension, Norfolk and Portsmouth, VA*. U.S. Department of Transportation Build America Bureau. https://www.transportation.gov/policy-initiatives/build-america/downtown-tunnelmidtown-tunnelmlk-extension-norfolk-and-portsmouth. Accessed July 18, 2017.
23. Tunnel Project Soaking Virginia. *The Washington Post*. October 26, 2015.
24. Nichols, K., and S. Belfield. Impact of New Tolls on Existing Roadway Facilities: Hampton Roads, Virginia, Experience. *Transportation Research Record: Journal of the Transportation Research Board*, 2016. 2597: 28–36.
25. SR 520 Funding Sources. Washington State Department of Transportation. http://www.wsdot.wa.gov/NR/rdonlyres/0A206E4F-3C0E-4BAF-82A2-63A931590E35/0/SR520Program_FundingSources.pdf. Accessed May 23, 2017.
26. *New Tacoma Narrows Bridge Project Case Study*. New York State Department of Transportation, New York State Thruway Authority, and Metropolitan Transportation Authority/Metro-North Railroad, 2008.
27. *Special Report 303: Equity of Evolving Transportation Finance Mechanisms*. Transportation Research Board of the National Academies, Washington, D.C., 2011.
28. Raimondo Signs RI Truck Tolls Plan into Law. *WPRI.com*. February 12, 2016.
29. ERC Revamping Late Fees, Billing Practices for Midtown, Downtown Tunnel Tolls. *The Virginian-Pilot*. June 2, 2017.
30. *NCHRP Report 689: Costs of Alternative Revenue-Generation Systems*. Transportation Research Board of the National Academies, Washington, D.C., 2011.

The Standing Committee on Revenue and Finance (ABE10) peer-reviewed this paper (18-00947).

Article

TRR
JOURNAL OF THE TRANSPORTATION RESEARCH BOARD

Transportation Research Record
2018, Vol. 2672(4) 65–76
© National Academy of Sciences:
Transportation Research Board 2018
Article reuse guidelines:
sagepub.com/journals-permissions
DOI: 10.1177/0361198118778928
journals.sagepub.com/home/trr
SAGE

Review of Handback Experience with Public–Private Partnerships

Qingbin Cui[1], Marcel Ham[2], Patrick DeCorla-Souza[3], and Emma Weaver[1]

Abstract

Public–Private Partnership (P3) handback experience is very limited in the U.S. and internationally. This paper analyzes the handback experience in three highway P3 projects: East-Link Bridge in Dublin, Ireland, Highway 4 (VT4 Järvenpää-Lahti) in Finland, and M4 Tollway in New South Wales, Australia. Two of the projects did not have any material handback clauses in their P3 contracts. Still, the handback processes and outcomes of all three projects are considered successful. The projects experienced a relatively smooth handback process, mainly because of very good working relationships between the contracting authority and the P3 concessionaire. There have not been significant technical quality concerns or problems in the years immediately following the handback. Other drivers for successful P3 handback include clear definition of handback requirements, sufficient incentives and protections, clear procedures and joint inspection processes, a collaborative approach, and workforce sustainment. These contractual requirements and financial incentives and protections are very beneficial as working relationships between P3 agencies and concessionaires will not always be excellent, which could lead to less smooth handback processes.

The U.S. public–private partnership (P3) experience is relatively recent in comparison to various other countries around the world. In the U.S., there have been no P3 highway projects which have reached the end of their contract term. This presents a gap in knowledge and experience regarding handback. Although there are few experiences internationally with completion of the handback process either, there are P3 projects which have reached this milestone.

In addition, several countries and international organizations have created guidance documents and reference guides for public–private partnerships and the handback process specifically. The sections relating to handback in these documents are often brief, but lay out the key elements of the handback process. The World Bank's *Public-Private Partnership: Reference Guide* states that the approach to the transition at the end of contract term should be clearly defined, and highlights defining asset quality as well as clearly specifying requirements and payment terms as critical handback elements (*1*). HM Treasury's *Standardisation of PF2 Contracts* echoes some of the World Bank's recommendations by stating that a contract should outline in detail the form of handback inspections, their timetable, consequences of failure, as well as the timing, procedure, and responsibility for acceptance (*2*). Other documents which provide some level of detail on the handback process include the United Nations' *A Guidebook on Public–Private Partnership in Infrastructure* (*3*) and the U.S. Department of Transportation's *Model Public-Private Partnership Core Toll Concession Contract Guide* (*4*). Both highlight critical elements to be included in handback requirements, with the UN identifying asset condition at the time of handover, as well as the obligations of the concessionaire, handover timeline, defect liability, procedures, and rights of the agency. The U.S. DOT adds that handback requirements should include valuation methodologies and inspection requirements based on asset condition and/or the required remaining useful life of an asset, and identifies common contract elements including the handback reserve account.

Other publications address the handback process, but do not go into a significant amount of detail. The Australian Government's *National Public-Private Partnership Guidelines* simply identifies the definition of handover quality as a key aspect of the contract (*5*). Similarly, South Africa's *National Treasury PPP Manual Module 6: Managing the PPP Agreement* defines key

[1]University of Maryland, College Park, MD
[2]IMG Rebel, Washington, D.C.
[3]Federal Highway Administration, Washington, D.C.

Corresponding Author:
Address correspondence to Qingbin Cui: cui@umd.edu

handback functions as implementing procedures associated with the transition, and acknowledges that this might involve the initiation of a new P3 agreement for post-handback operation and maintenance, but provides no additional detail (*6*).

Various papers review current and best practices in handback specifications, as well as residual value risk in P3 projects. Startin discusses current and best practice for handback provisions including distinguishing between elements with long and medium service life in inspection procedures, and residual life requirements (*7*). Startin also defines good handback requirements as appropriate and enforceable, and states that additional research is needed on the successful application of handback reserve funds. Yuan et al. identify residual value risk, or the risk that an asset will not have the remaining useful life the public agency expected at contract expiration, as an important issue in relation to P3 projects through surveys of experts, and acknowledge that differences of opinion exist on the best method for managing this risk (*8*). They also identify six perspectives from which residual value risk should be managed, including deterioration of maintainability and decline of operability. In an empirical study of P3 projects in Canada, Yuan and Li found that service life of pavements as well as the remaining useful life upon contract expiration under P3 delivery is higher than under conventional delivery (*9*). They suggest that this superior performance is in part attributable to handback requirements, as well as a performance-based payment mechanism.

Although there are a range of documents internationally that attempt to identify key contract terms and elements of the handback process, there appears to be a significant knowledge gap in terms of practical knowledge and experience with the handback process; the World Bank Reference Guide specifically acknowledges the limited practical experience and guidance in handback requirements. In most cases, documents drawing on practical experience included brief lessons learned and not in-depth analyses of the critical elements of the P3 handback process.

This paper is intended to respond to the gap in knowledge and experience regarding handback. This study begins with a review of criteria for selection of case studies followed by an evaluation of the P3 handback practices in each case. Next, the following section outlines drivers for successful handback that were developed as a result of these evaluations, which is finally followed by recommendations for successful practices in the U.S.

Analytical Framework

Several critical elements of the P3 handback were identified through a review of existing knowledge on the P3 handback process. These critical elements include the following.

- **Handback requirements**—clear identification and definition of handback requirements within the contract were identified as a critical step in most documentation. However, specific recommendations and best practices were not provided in most guidance.
- **Monitoring/inspection procedures**—the presence and clear definition of monitoring and inspection procedures are also critical to project handback. As with handback requirements, specifics related to the timing and focus of inspections were not provided in most guidance.
- **Financial incentives/securities**—financial incentives and security specifically related to the handback process will be critical in any study of the handback process. However, limited guidance was present in the reviewed documents.

Consideration was given to these elements as they relate to the identified case studies. In addition to these key elements identified in the review of existing knowledge on the P3 handback process, more recent contracts contain detailed handback elements. Table 1 details key handback elements identified in these more recent handback clauses for reference; some of these elements were identified in the reviewed case studies as well. The handback elements identified in recent P3 contracts echo those found in the literature and generally fall into the three categories identified above: (1) handback requirements, (2) monitoring/inspection procedures, and (3) financial incentives/securities. However, the key handback elements identified in recent contracts provide more detail than was present in the review of existing knowledge on the P3 handback process. The elements identified in Table 1 are based on recent P3 contracts, contracts executed within the last four years. These contracts represent a range of project types as well as a range of geographies, and were selected to ensure coverage and representativeness of the market.

A handback clause can be evaluated as a success based on the outcome as well as the handback process. The handback outcome is successful if the returned asset meets P3 Agreement requirements at no additional cost to the public agency. The following criterion will be used in this paper:

- The asset was returned meeting or exceeding requirements set forth in the P3 Agreement at no additional cost to the contracting agency.

As many early P3 Agreements do not define handback requirements, the handback outcome must be evaluated

Table 1. Summary of Key Handback Elements

Key element	Description
Inspection requirements	A minimum number of inspections are required, on or by a date that is specified in the contract. Often these are required to be joint inspections between the public and private sector, or involve a third party.
Minimum residual life requirements	Minimum criteria outlined in a requirements table that must be met for each element or asset included in a project. This list is typically comprehensive.
Handback plan	This plan is developed by the concessionaire and submitted for approval prior to the handback process. It details all required renewal work, inspections, key steps in the transition and proposed training. This plan is typically required to be updated on a yearly basis.
O&M training session	The private is required to provide the public-sector employees with training on the operations and maintenance of the facility. This training must be completed by some previously defined time prior to the end of the project term. In addition to this, the private company must make personnel available for public-sector consultation on O&M and repair work for some period after the project term expires.
Handback reserve account	The concessionaire is required to establish and fund a handback reserve account. Deposits into this account are required quarterly.
Spare parts & tools	The concessionaire is required to provide the public sector with all spare parts and tools necessary for operations and maintenance of the facility.
Residual life calculation methods	The concessionaire is required to propose methods for calculating Residual Life. This helps to avoid the use of outdated standards and practices.
Final handback acceptance	A certification that releases the concessionaire from the P3 contract and final payment if applicable.

through other subjective criteria, which for the purpose of this paper are defined as follows:

- The asset was returned in equal or better condition as compared with an asset of similar nature, age, and geographical location.
- The public authority was not required to make capital expenditures on repairs or upgrades to ensure the asset is in working condition going forward.

To be considered a success, the handback process should be evaluated by the extent to which it followed the procedures defined in the P3 Agreement. The following criterion that will be used in this paper:

- The handback procedures were implemented as defined in the P3 Agreement including following procedures for inspection, handover, financial guarantees, and warranty provisions.

As many early P3 Agreements do not define handback procedures, the handback process must be evaluated through other subjective criteria, which for the purpose of this paper are defined as follows:

- The handback process resulted in no added transactions cost to the public agency, including costs related to delays, inspections, litigation, and other material expenses beyond what is reasonably expected.
- The returned asset was made available for use throughout the handback process without limiting service or performance.
- The handback process avoided conflict between public agency and concessionaire, and resulted in no litigation or need for mediation between parties.

Particularly in cases where handback procedures are not clearly defined, an external audit may be helpful in documenting lessons learned, and ensuring that outcomes are in the public interest. One reviewed case study, the M4 Tollway in New South Wales (NSW), Australia incorporated an external audit of the handback process; however, it was not utilized in other cases.

Various methods were utilized to identify qualified case studies for this research including web-searches, outreach, and the existing knowledge of the project team. Outreach was conducted to various P3 offices and organizations worldwide to identify candidate case studies that have completed handback or are currently undergoing the handback process. In most cases, P3 projects, particularly highway projects, had either not yet reached handback, were terminated prior to the contract end date or in cases where handback would have been reached the concessionaires' contracts were extended. These factors increased the difficulty of identifying qualified candidates for this study. Once qualified case studies were identified, all available project documentation and existing publications on the project were collected for review. After documentation was reviewed, project contacts and other potential interviewees were identified and contacted to set up interviews. These interviews were to fill the gaps in available knowledge and provide additional context.

Although projects were identified worldwide, the priority was placed on highway projects from English-speaking countries for ease of data collection. Two projects with superior data availability were identified in English-speaking countries. These projects are the M4 Tollway in New South Wales, Australia, and the East-Link Bridge in Dublin, Ireland. However, no other projects were identified that met the primary selection criteria, completed handback, and had data available in an English-speaking country. Therefore, the Highway 4 (VT4 Järvenpää-Lahti) project in Finland was selected, even though the official languages of Finland are Finnish and Swedish. Project documentation for this project, including the original contract, as well as case studies conducted after the fact, was available in English. English-speaking interviewees were identified for this project as well.

P3 Handback Case Studies

East-Link Bridge: Ireland

The East-Link Toll Bridge (The Bridge) spans the River Liffey in Dublin, Ireland connecting North Wall to Ringsend. The Bridge was completed in 1984 and replaced the ferry service which had been carrying passengers across the river since 1665. As of 2016, the bridge averaged 16,000 vehicles per day and charges tolls for all trucks (2.6–5.2 Eur) or cars (1.75 Eur) using cash and electronic transponders. The East-Link Bridge was one of the first P3 projects procured in Ireland. Currently, there are fourteen P3 projects in the transport sector, and many more in various other sectors.

The concession was awarded through a Design-Build-Finance-Operate-Maintain (DBFOM) contract to National Toll Roads (NTR), an Irish limited company who designs, builds, finances, and operates toll roads in Ireland. NTR built and operated the Bridge through a special purpose vehicle (SPV) named Eastlink LTD. The concession was a toll-based revenue contract, with proceeds from the fares allocated to the City Council (17%), Dublin Port Company (25%), and NTR (58%) (*10*). While the Dublin Port Company did not directly invest, in exchange for the disposal of land and limited berthage, they were given the right to profits from the toll scheme until contract end, or the date on which the building costs were finally discharged (*11*). As a result of this profit-sharing mechanism, revenue risk is proportionally shared among the three parties. In 2010, NTR sold the East-Link Bridge to the Dutch Investment Fund as part of a package of assets for EUR 50 mil, with O&M transferred to Egis Road Operations.

The contract expired on December 31, 2015 and on January 1, 2016, East-Link transferred tolling and maintenance responsibilities to Ringsend Toll Bridge, which is wholly owned by the Dublin City Council (DCC). The original concession agreement had no handback requirements or procedures and the document adopted an approach that the handback should be viewed as the concession rights simply ending rather than returning an asset. Throughout the process, the legal team advised the City Council to develop a handback agreement with detailed procedures and penalties. The only challenge in the handback process was the transfer of employees and staff of Eastlink from private employment to public employment. In general, the transfer of staff is a complex and sensitive issue, and should be viewed as a risk.

The success of the East-Link handback is attributed to the close relationship between Eastlink and DCC. Under terms of the original agreement, the Board of Eastlink was required to include one administrative official and two elected members of the DCC. Given this unique arrangement, regular interaction between parties during the life of the agreement was guaranteed, allowing the DCC to confirm maintenance plans and be notified of any issue well in advance of handback. This strategy allowed members of the Board to review maintenance budgets, oversee specifications on maintenance, including inspections, and receive maintenance and inspection reports throughout the course of the concession. This access to information and knowledge of decisions regarding maintenance provided a sense of comfort to the DCC that material defects would not be found. The East-Link Bridge was considered a success despite a lack of formal handback language in the concession agreement. The handback procedure more closely followed an end of a service agreement with a handshake agreement to meet performance conditions. At multiple stages lawyers recommended to formalize a handback process and procedure; however, both parties felt it was not needed. In reflection, DCC indicated that while successful, a contract with greater detail on the handback procedure and requirements would have been useful.

- **Outcome:** On January 1, 2016, responsibility for collecting tolls and operating the bridge was transferred from Eastlink to the Ringsend Toll Bridge (DCC). A post-concession inspection revealed only minor issues consistent with what would be expected from a similar bridge.
- **Process:** The handback process occurred without incident with procedures in place to ensure operability was not impacted as the bridge was

transferred to the DCC. Like the M4 Tollway, the success of the East-Link Bridge handback process was based on the relationship between Eastlink and the DCC. The DCC was part of the Eastlink board of directors, and therefore had a vested interest in the ongoing operations of Eastlink throughout the term of the contract. Throughout the life of the contract, and including handback, the DCC monitored maintenance regimes and was able to interact with Eastlink to ensure no issues arose at handback.

Highway 4 (VT4 Järvenpää-Lahti): Finland

The Highway 4 project in Finland (European highway designation E-75) was the first roadway project in the country to be procured as a P3 using a DBFOM model, with a shadow toll as the payment mechanism, where the concessionaire was paid based on annual traffic volumes. The project included a 70-kilometer motorway, 88 new bridges, 8.5 kilometers of noise barriers and 130 kilometers of moose fence (*12*). Construction of the project lasted 2.5 years, less than the originally anticipated 3 years and significantly less than expected if procured using traditional methods. The total period of the concession was 15 years. The originally estimated cost of the project was 238 million Euros; however, the actual project cost was only 234 million Euros. The agreement between the Finnish Transport Agency (FTA) and the SPV, Tieyhtiö Nelostie Oy, was signed on March 19, 1997 (*12*).

Handover occurred at midnight on August 30, 2012, and was considered an overall success by both the public and private partners, as was the project as a whole. The handover process began 3 years prior to the end of the project term; the process began when the concessionaire submitted a request to the FTA, per contract requirements. Generally, the facility was required to be in the same condition as other motorways of similar age and class in Finland, per the original contract requirements, with specific criteria at handback also included. These criteria included maximum rut depth, groundwater protection measurements, and requirements related to the functionality and cleanliness of the drainage systems among others. A preliminary handover inspection occurred 5 months prior to contract expiration, in the spring of 2012, with the final transfer occurring after this in two parts.

In addition, the SPV was required to submit all information required for the public sector to estimate the year of next repair, per the original contract requirements. The contract required a two-year warranty period during which the SPV was responsible for any larger than expected, or extra high maintenance costs. However, this threshold was not specifically determined, nor were there any issues during this period. Overall, the relationship between the parties remained good throughout the contract and handback periods. The Highway 4 P3 Agreement included detailed requirements on the condition of the highway at transfer, as well as defining a three-year timeline to execute the handback. The FTA reviewed the concession after maturity, and recommended the Highway 4 P3 serve as a guideline for future P3 handbacks.

The Highway 4 project in Finland (European highway designation E-75), was the first roadway project in the country to be procured as a P3. In addition to the Highway 4 project, Finland has pursued three other P3 projects, two of which have not reached handback. Procurement of the third P3 project was suspended in 2011 because of higher than expected bid prices.

- **Outcome:** At handback the highway met all requirements as defined in the original P3 agreement. The agreement explicitly required the road to be in a similar condition to any other 15-year-old motorway in Finland (*12*). The road was also subject to requirements relating to maximum rut depth, functioning drainage system, groundwater protection measures, reflectivity of signage, road lighting, condition of noise barriers, as well as other technical requirements. The successful handback is also attributed to a relatively short concession term, given the fact that a typical design life of 40 years was used for highways in European countries (*13*).
- **Process:** The handback followed a three-year handback timeline which was executed at Tieyhtiö Nelostie Oy's request. During the handback process, the FTA was able to perform their own inspections, and joint meetings were held routinely to ensure the handback process met timing and condition requirements. Following the transfer inspection, Tieyhtiö Nelostie Oy provided a two-year warranty period ensuring there were no added extraordinary costs to the public agency.

M4 Motorway: NSW, Australia

The M4 Tollway in NSW, Australia was one of the first P3s in any sector initiated by the NSW government, and was the first Australian road concession to reach handback. The project included construction of a missing section of the M4 Motorway from Mays Hill to Prospect, and improvements to lane configurations in other sections; the original sections of the M4 Motorway were built in the 1970s and 1980s. The M4 Tollway was scheduled to open in 1993, but the project was nine months ahead of schedule and was opened to the public in May of 1992; both the cost and schedule of the project were below original estimates. The total length of the M4 Motorway runs from Strathfield to Lapston and is one of Sydney's main arteries. The missing section and improvements were constructed through a 20-year, DBFOM toll concession with the concessionaire, who was Statewide Roads (SWR). The project owner is the NSW Roads and Traffic Authority (RTA), which is now part of the Roads and Maritime Services, a NSW transportation agency. The total capital cost of the project was A$246 million (*14*). Throughout the course of the concession, SWR underwent several ownership changes. In 2007, just before handback Transurban became majority shareholder. During the concession period, the M4 Motorway consisted of both tolled and non-tolled section of the roadway, with the tolls paying for the privately built portions of roadway. During the concession period, the government changed hands from the Liberal Party, under which the project was initiated, to the Labor Party which campaigned on removing tolls on several roadways; the Labor Party government was elected in 1995. It was discovered that removing the tolls would require the government to compensate the concessionaire a large sum if this action was undertaken. Instead, the government of NSW instated a "Cashback Scheme" that allowed drivers of privately registered vehicles to claim a refund on tolls paid. Ultimately, this program repaid the users of the M4 A$255 million; a total of A$1.2 billion was paid by motorists over the course of the concession period (*14*). Ultimately, to fulfill their promise the Labor Party government removed all tolls on the M4 Motorway once the facility was handed back to the public sector.

Handback of the M4 Tollway occurred at midnight on February 15, 2010. The original project deed between SWR and RTA did not include a specific handback provision. Most significantly, it did not specify how to determine if the roadway was in satisfactory condition at handback, how to determine what repairs were needed, how to cover costs for repairs not made, and how maintenance requirements should change with evolving standards (*15*). The maintenance requirements that were included in the contract were based on 1990 standards. Despite these challenges, an audit of the handback process conducted by the Auditor-General of New South Wales indicated that the facility was expected to be in satisfactory condition at handback, thanks in large part to good management by both RTA and SWR, and goodwill between the parties.

The handback process began in November 2008 when a governance structure was set up to oversee contract closure. The governance structure included both parties, with members of the management of SWR and RTA. These parties governed all items related to handback, including the steering committee which dealt with all handback details. The steering committee overseeing all components of handback consisted of three subcommittees: commercial and legal; asset management; traffic and safety management. Each subcommittee had its own unique objectives related to handback. The outcomes of processes related to handback and final decisions that were made were formalized in a project closure deed to make the process official (*16*). Throughout the handback process bi-monthly meetings were held between handback teams at SWR and RTA to update RTA on progress and clarify any new or outstanding items. These handback teams, comprising representatives from both SWR and RTA, over the two-year period defined all quality and standard requirements at handback, as well as other handback processes.

The P3 Agreement for the M4 Tollway did not clearly define the handback process, maintenance requirements, or minimum asset conditions. SWR was only required to return the motorway in "satisfactory condition." The handback process was defined on an ad hoc basis outside the P3 agreement through the establishment of a steering committee, and later formalized in the project closure deed. Chung and Hensher describe that the ability to invent a handback process during the course of the project was a successful strategy that helped lead to a successful handback (*14*). However, the Auditor-General of NSW recommended that future project deeds include appropriate testing and inspection programs to determine repairs needed for satisfactory asset condition at handback (*15*).

As mentioned, the M4 Tollway was one of the first P3s in any sector initiated by the NSW government, and was the first Australian road concession to reach handback. Since completion of the M4 Tollway, the NSW Roads & Maritime Services has awarded eight P3 contracts, as well as many others in various sectors. Achterstraat notes that the audit of the M4 Tollway project was conducted in part so that lessons learned from the project could be incorporated into future P3 projects and handback clauses (*15*).

- **Outcome:** The tollway was returned in satisfactory condition with minimal financial burden to the authority. Post-concession inspections found a defect in one of the bridges; however, the defect was a result of components installed before the beginning of the concession, and not the liability of SWR (Transurban).
- **Process:** The steering committee was formed in 2008 to define and formalize the handback process and requirements. SWR and the New South Wales Government met on a bi-monthly basis to clarify roles and address any issues that occurred

Table 2. Handback Processes and Outcomes Evaluation

Project	East-link bridge	Highway 4	M4 tollway
Clear handback requirements in P3 agreement	No	Yes	No
Handback outcome met criteria for success	Yes—bridge was returned in satisfactory condition at no additional cost to the City of Dublin	Yes—highway met all criteria at no additional cost to the Finnish Transport Agency	Yes—the tollway met conditions of a similar asset, the only added capex being the repair of a pre-concession bridge defect
Clear handback procedures in P3 agreement	No	Yes	No
Handback process met criteria for success	Yes—no conflict, cost, or performance impact	Yes—process followed handback timeline and requirements	Yes—the process was free from conflict, material cost, or performance impact

during handback. The relationship and goodwill between stakeholders was a key driver in the success of the M4 handback. Transurban, the majority owner of SWR, had multiple concessions with the Australian government, and was motivated to preserve the relationship and comply with any requirements to meet a "satisfactory condition." The handback took place over a two-year period without incident, cost, or impact on performance of the tollway.

Table 2 provides a high-level evaluation of the handback processes and outcomes in each of the cases reviewed above.

Evaluation and Takeaways of Handback Experiences

All three case studies showed the handback process being executed without incident or added cost, and with the asset returned in satisfactory condition to the public authority. However, as highlighted by the M4 and East-Link projects, satisfactory outcomes were in part driven by strong relationships between stakeholders. In the East-Link and M4 case studies, incentives were aligned through strong relationships between the concessionaire and public agency. Further, the concessions offered no recourse for the public agency if assets were not returned in satisfactory condition. This placed significant financial risk on the public authority, who may have had to repair the asset so that it met performance standards. Based on the case studies and literature review, key takeaways are highlighted below for effectively mitigating handback risk and improving the chances of a successful transfer:

- **A lack of clear handback requirements can lead to an ineffective handback process.** Vague requirements, such as a requirement that the asset be returned in "satisfactory condition," can lead to conflicting interpretations on a satisfactory benchmark. Whereas the various case studies presented in this paper do not convincingly demonstrate that the outcome of the handback is optimal by objective standards, they do show that even without clear handback requirements, parties may consider the handback process successful.
- **Clear procedures and a (joint) inspection process are critical success factors:** Respondents confirm that successful handback processes clearly define the roles and responsibilities of the inspection procedure. This includes establishing the timing of inspections, benchmarks for asset quality, and procedures for incorporating inspection findings into maintenance plans. To maximize the chances of success, inspections should be conducted jointly to ensure a transparent and common understanding of the results.
- **Sufficient financial incentive and protection are beneficial:** Respondents representing the various case studies in this report indicated that they would have benefited from financial incentive and protection mechanisms. Such mechanisms effectively limit the threat of perverse behavior and profit maximization of the concessionaire. Moreover, through establishing maintenance benchmarks supported by maintenance reserves, letters of credit, or surety bonds the public agency is protected financially, should the asset require capital investment or repairs.
- **Flexibility regarding handback requirements is required:** Technical requirements and contract provisions tend to be incomplete owing to a changing environment. This is especially important for long-term contracts where specification of

condition and performance criteria possibly becomes inadequate or impractical. As shown in all three projects cases, no agency worked with detailed handback requirements. Instead, flexible contract provision, effective organizational structure, and collaborative effort among stakeholders led to satisfactory handback.

- **Short contract durations reduce the challenges of changing requirements:** P3 contract terms generally are 30 years or more. Such long contract terms obviously increase the likelihood of significant changes in technical and performance standards. It is likely that the relatively short term of the P3 contract for the Highway 4 project contributed to the successful handback process. In general, shorter contract terms will reduce the challenges of the need to change handback requirements, as well as technical and performance standards in general. This, however, needs to be balanced with life cycle optimization considerations, where one or multiple rehabilitation programs are generally required.
- **A collaborative approach contributes to allow for necessary flexibility:** As the case studies have shown, a strong relationship between concessionaire and public agency enhances the chances of handback success, even in the absence of detailed requirements and procedures. In general, international P3 contracts seem to be less legalistic and focus more on collaboration than U.S. P3 contracts. Further research is needed to confirm to what extent a more collaborative P3 approach would be applicable in the U.S.
- **Workforce transition upon handback can be complicated:** As shown in the East-Link example, one often ignored aspect of the handback process is the transition of any employees from the private concessionaire to a new operator, be it the public agency or new private sector operator. A successful handback secures continuity in operations by integrating existing employees into the new operator's workforce, while at the same time maintaining operations and service quality of the asset for users. The public agency needs to capture and consolidate private sector knowledge for successful handback, and manage the transition to a new operator. However, it should be noted that in general the transfer of staff involves significant complexity and creates uncertainty for both the public agency and the concessionaire.

Successful Practices

Based on the international case studies, as well as the evolution of P3 contracts, there are clearly identifiable successful international practices that could be incorporated in future U.S. P3 contracts. A review of several recent P3 contracts in the U.S. and internationally, shown in Table 3, demonstrates that these successful practices have been incorporated into the more recent generation of P3 contracts.

Handback Requirements: Standards

In drafting P3 contracts and legal documents, public agencies and concessionaires ensure handback procedures and requirements are clearly defined. This includes establishing clear maintenance and inspection requirements as well as defining the handback process. Detailed maintenance requirements will ensure the concessionaire is maintaining the asset to an acceptable level of performance rather than maximizing profit. However, it is equally important to incorporate flexibility in the P3 contract by defining the requirements in functional, output-based terms rather than prescriptive, input-based terms and allowing for changes to the requirements so that the acceptable state of practice at a given time is applied, not one that is outdated and potentially detrimental to the public interest. This is relevant for all performance requirements, throughout the life of the contract, not just at handback. Recognizing this challenge, public agencies should balance the trade-off between strict contract provisions and flexibility for negotiation and inspection to ensure no requirements become impracticable or inexecutable because of technological advancement or policy changes.

Handback Requirements: Process

The P3 agreement also defines the handback process including when the process should begin, roles, and responsibilities. Clearly establishing responsibilities will ensure handback occurs according to schedule and without conflict. An analysis of a selection of more recent and ongoing P3 projects, including the M4/M5 WestConnex project in Australia, and the I-4 Ultimate and I-77 Express Lanes projects in the U.S. all include clearly defined handback requirements, initiation dates for the handback process, and roles and responsibilities. This suggests that these recommended improvements have been incorporated into more recent P3 contracts. Additional best practices, including requirements to turn over key materials, are often included in more recent handback provisions such as those mentioned above. These key handback elements are shown in Table 3.

Monitoring/Inspection Procedures

P3 contracts develop a Life-Cycle Maintenance Plan, defining requirements and inspection procedures

Table 3. A Comparison of P3 Handback Provisions

P3 project	Eastlink	M4 tollway	Highway 4	WestConnex	I-4 ultimate	I-77 expressway	SH 183 managed lane
Country	Ireland	Australia	Finland	Australia	U.S.	U.S.	U.S.
Contract date	1982	1990	1997	2015	2014	2014	2014
Handback defined (contract)	None	None	Not defined	Handback requirements defined	Handback requirements defined	Handback requirements defined	Handback requirements defined
Initiation of handback procedure	Termination of contract	Not defined	Inspection up to 3 years before termination of contract	3 years prior to expiry date	4 years prior to the termination of the contract	5 years prior to the end of the contract term	5 years prior to the end of the contract term
Number of inspection rounds	One at handback	Not defined	One transfer inspection	Two (3 years prior and 18 months prior to expiry date)	Annually	Annually	3 (60 months prior, 18 months prior, and 30 days prior to end of the term)
Responsibility for inspections	Shared	Not defined	Shared	Shared	FDOT	NCDOT & developer	Developer, with TXDOT able to witness as desired
Financial mechanism—type	Contingency fund provision was removed mid-contract	None	Maintenance cost collected from concessionaire	Escrow account established after the initial inspection	Handback requirement reserve account (or L/C) established 4 years before termination	Handback requirement reserve account (or L/C) established 5 years before termination	None
Financial mechanism—sizing	None	None	Maintenance cost	40% of the total estimated cost of renewal work	Equal to the renewal amount as estimated following each inspection of the project	Equal to the renewal amount as estimated following each inspection of the project + 10% contingency	None
Handback requirements vs. regular requirements	Same	Unknown	Unknown	Regular inspections	Handback requirements detailed in technical requirements	Handback requirements detailed in technical requirements	Handback requirements detailed in technical requirements

(continued)

Table 3. (continued)

P3 project	Eastlink	M4 tollway	Highway 4	WestConnex	I-4 ultimate	I-77 expressway	SH 183 managed lane
Post-handback guarantee	None	None	Concessionaire guarantees that no significant maintenance beyond regular maintenance on similar roads would be necessary for a period of 2 years from transfer inspection. Concessionaire is responsible for any extra maintenance costs. Final receivables of the contract are paid out only after this period.	Within 60 business days of the expiry date RMS may identify assets which don't meet "residual life requirements" to be addressed by the project company	None	None	None
Residual life (RL) requirements	None	None	The road must be returned in a similar form of the same class and age (15 years)	Detailed in the additional requirements	No RL requirements but with detailed conditional criteria at handback	Detailed RL requirements for every asset type	Detailed RL requirements for every asset type
Level of detail of handback requirements	None	None	High	High	High	High	High

throughout the life of the agreement. This insures the public agency and concessionaire are both aware of the asset condition and can take steps to mitigate issues, such as increasing the maintenance reserve, well in advance of handback. The inspections take place in a joint fashion, with roles clearly defined, to ensure both parties are in agreement with any findings and understand the financial ramifications. By clearly identifying monitoring and inspection procedures and utilizing a joint approach, the financial risk from unexpected repairs at handback will be minimized. The M4/M5 WestConnex project in Australia and the I-4 Ultimate and I-77 Express Lanes projects in the U.S. all include clearly defined inspection procedures, and requirements detailing their frequency. In addition, the M4/M5 WestConnex and I-77 Express Lanes projects require joint inspections, or shared inspection responsibilities, demonstrating that these improvements have been incorporated into more recent projects.

Financial Incentives/Securities

Finally, should the asset be returned to the public agency in an unacceptable form or require extraordinary maintenance following handback, the public agency is protected financially. This is complicated, as the concession is typically run by an SPV who may dissolve after the concession matures and no longer has capital. Through financial incentives and protection including a handback reserve account, surety bonds, and/or letters of credit that extend beyond maturity of the contract, a public agency can be guaranteed that their financial risk at handback is minimal. All recent projects detailed in Table 3 include some form of Escrow or handback reserve account, and specific requirements regarding its sizing. This demonstrates that the suggested improvements have been incorporated into more recent P3 contracts.

Conclusion

Analysis of international experiences to date identifies clear drivers of handback success and successful practices. Admittedly, the M4 and East-Link case studies relied significantly on strong relationships to achieve success; however, absent these traits, the P3s could have exposed the public agency to financial risk given a lack of recourse and clear performance standards in the legal documentation. Through focusing on the drivers of success and handback best practices, a public agency can safeguard itself from the financial risks associated with a returned asset while ensuring that the P3 is successful. There is evidence that some public agencies have begun this work already, demonstrating acceptance of these drivers of success and handback best practices.

Acknowledgments

This research was supported by the U.S. Department of Transportation Volpe Center and Federal Highway Administration. We would like to show our gratitude to all participants who provided insights and data that greatly assisted the research. We are also grateful to the reviewers for their constructive comments on an earlier version of the manuscript.

Author Contributions

The authors confirm contribution to the paper as follows: study conception and design: Patrick DeCorla-Souza and Qingbin Cui; data collection: Emma Weaver and Marcel Ham; analysis and interpretation of results: Marcel Ham and Qingbin Cui; draft manuscript preparation: Marcel Ham and Emma Weaver. All authors reviewed the results and approved the final version of the manuscript.

References

1. World Bank. *Public-Private Partnerships: Reference Guide Version 2.0.* World Bank Group, Washington, D.C., 2014.
2. HM Treasury. *Standardisation of PF2 Contracts.* HM Treasury, London, United Kingdom, 2012.
3. Quium, A. *A Guidebook on Public–Private Partnership in Infrastructure.* United Nations Economic and Social Commission for Asia and the Pacific (ESCAP), Bangkok, 2011.
4. FHWA. *Model Public-Private Partnerships Core Toll Concessions Contract Guide.* Federal Highway Administration, Washington, D.C., 2014.
5. Australian Government, Department of Infrastructure and Regional Development. *National Public Private Partnership Guidelines Volume 2: Practitioners' Guide.* Commonwealth of Australia, 2008.
6. South Africa, National Treasury. *National Treasury PPP Manual Module 6: Managing the PPP Agreement.* South Africa, National Treasury, Johannesburg, 2004.
7. Startin, J. Value from Public-Private Partnerships: Balancing Prescriptive and Performance Specifications from Design to Handback. *TR News*, No. 274, 2011.
8. Yuan, J., A. P. Chan, W. Xiong, M. J. Skibniewski, and Q. Li. The Perception of Residual Value Risk in Public Private Partnership Projects: A Critical Review. *Journal of Management in Engineering*, Vol. 31, No. 3, 2015, p. 04014041.
9. Yuan, X.-X., and Y. Li. *Understanding the Residual Value Risks of Highway Pavements in Public-Private Partnerships: Part I—An Empirical Study.* Ryerson Institute for Infrastructure Innovation, Ontario, Canada, 2016.
10. RTÉ. AA Critical of Keeping Toll on East-Link Bridge. *RTÉ*, 31 December 2015–9 April 2017.
11. Dublin Port Company. *Growing with Dublin and Ireland.* Dublin Port Company Annual Report 2015, Dublin, Ireland, 2015.

12. Makinen, S., P. Petajaniemi, and P. Pakkala. Results of Finland's First Concluding Public-Private Partnership Road Project. Prestented at Transportation Research Board 93rd Annual Meeting (No. 14-0589), Washington, D.C., 2014.
13. Hall, K., D. Dawood, S. Vanikar, R. Tally, Jr., T. Cackler, A. Correa, P. Deem, J. Duit, G. Geary, A. Gisi, and A. Hanna. *Long-Life Concrete Pavements in Europe and Canada (No. FHWA-PL-07-027)*. Federal Highway Administration, Washington, D.C., 2007.
14. Chung, D., and D. Hensher. Risk Management in Public–Private Partnerships. *Australian Accounting Review*, Vol. 25, No. 1, 2015, pp. 13–27.
15. Achterstraat, P. *Auditor-General's Report Performance Audit: Handback of the M4 Tollway*. Audit Office of New South Wales, Sydney, Australia, 2009.
16. Chung, D. Risks, Challenges and Value for Money of Public–Private Partnership. *Financial Accountability & Management*, Vol. 32, No. 4, 2016, pp. 448–468.

The Standing Committee on Revenue and Finance (ABE10) peer-reviewed this paper (18-00684).

The opinions, findings, conclusion, and recommendations expressed in this paper are those of the authors and do not necessarily reflect the view of the FHWA, Volple Center, or the US government.

Article

TRR
JOURNAL OF THE TRANSPORTATION RESEARCH BOARD

Transportation Research Record
2018, Vol. 2672(4) 77–89
© National Academy of Sciences:
Transportation Research Board 2018
Article reuse guidelines:
sagepub.com/journals-permissions
DOI: 10.1177/0361198118775875
journals.sagepub.com/home/trr
SAGE

Microeconomic Model for Designing Public Transit Incentive Programs

Yanshuo Sun[1] **and Lei Zhang**[1]

Abstract
The purpose of this study is to propose a microeconomic model for optimizing a government-sponsored transit incentive program, where a rider receives a certain monetary reward once a transit trip is completed. The demand pattern considered is many-to-many, which means riders travel from multiple origins to multiple destinations. Travelers are grouped by their socio-demographic characteristics and incentives can be customized to each rider group. Analytical relations are derived from the optimality conditions of the optimization problem and numerical results are conducted to illustrate some important findings. It is found that (a) the eligibility of a certain rider group for transit incentives depends on the magnitude of the cost of public funds, and (b) an increasing profit level diminishes the social welfare. The effect of cross-subsidization among multiple origin–destination pairs is also identified and analyzed.

In a densely populated urban environment, driving is discouraged almost ubiquitously because of its substantial negative externalities, such as congestion, air pollution, emissions, and safety hazards. Since the public transportation industry exhibits considerable economies of scale, the cost per passenger-mile is relatively low. In other words, public transportation can be very cost effective, energy efficient, and environmental friendly, if the transit patronage is sufficiently high. Therefore, governments at all levels, universities, and major companies employ different strategies and design various programs to attract travelers to public transportation from other modes (*1–3*). This paper examines the optimal design of a transit incentive program, where a rider receives a certain monetary reward once a transit trip is completed.

In the empirical literature, extensive real-world studies have examined the effects of various transit benefit programs, such as a discounted transit pass program. Those studies show that, in general, transit benefit programs can increase the transit ridership and can also be financially justifiable (*2*). In the theoretical literature, there is still no consensus on transit subsidization or incentivization, and researchers draw completely different conclusions. For example, while most studies cite the "Mohring effect" (*4*) as the rationale for subsidizing public transit services, Van Reeven (*5*) argues this effect does not constitute a valid justification for subsidization. Both Savage and Small (*6*) and Basso and Jara-Díaz (*7*) refute the conclusion of Van Reeven (*5*) but with different models. Even if it is decided that subsidization or incentivization is a necessity, new problems, such as to whom subsidies or incentives should be provided and how they should be distributed, should be solved. A few studies (*8–10*) have recently tackled the optimal design of transit subsidization schemes and the following issues have not been addressed sufficiently:

(1) Most studies assume the transit operator to be the only recipient of incentives.
(2) Demand modeling is unrealistically simple, e.g., all travelers travel from the same origin to the same destination.
(3) Rider heterogeneity is largely ignored, e.g., all travelers have the same demand elasticity and the same value of time.

In practice, many government-sponsored or employer-sponsored transit programs transfer financial incentives directly to transit riders or employees. Transit travelers have different trip origins as well as destinations, and therefore quite different trip distances and costs. Since travel volumes are not uniformly distributed over space,

[1]Department of Civil and Environmental Engineering, University of Maryland, College Park, MD

Corresponding Author:
Address correspondence to Lei Zhang: lei@umd.edu

the transit operator should ensure the spatial peak demand is met, which gives rise to the effect of "cross-subsidization" (this will be discussed later in this paper). More importantly, it is well known that transit riders are not alike and might behave differently—even within the same transit service—due to socio-demographic differences. For transit-dependent riders, transit is their only travel option; others have the privilege of choosing among various transportation modes. Trip purpose also significantly influences riders' impatience (which could be measured with value of time) with transit services, in addition to other factors, such as income level. Therefore, rider heterogeneity should be explicitly considered in designing any transit incentive program, and incentives tailored to the target groups can bring the most overall benefit.

The focus of this paper is to optimize a government-sponsored transit incentive program. The demand pattern considered is many-to-many, which means riders travel from multiple origins to multiple destinations. Travelers are grouped by their socio-demographic characteristics and incentives can be customized to each rider group. In addition to determining incentive amounts, the government determines the fare and headway limits for the operator to follow, with an objective of social welfare maximization. The operator, which is aware of the policy constraints enforced by the government, optimizes its operations to maximize its profits.

The remainder of this paper is organized as follows. The next section reviews both empirical studies and theoretical work on transit incentivization, after which a microeconomic model for optimizing transit incentives is proposed. Then, important analytical relations are derived from the optimality conditions of the proposed problem. Extensive numerical analyses are conducted to demonstrate the capabilities of the model. This paper closes with major findings and prospective research directions.

Literature Review

Empirical Studies

By U.S. federal tax law, tax-free transit benefits can be offered to employees. Federal employees in the National Capital Region are eligible to receive the federal transit benefit if they commute to and from work using public transportation (*11*). The Maryland Transit Administration has a program, "Commuter Choice Maryland" (*12*), through which transit benefits are distributed. There are three options: (a) the employer-supported option, (b) the pre-tax salary deduction option, and (c) a combination of (a) and (b). Motivations for employers to support such transit benefit programs include receiving a federal and a state tax deduction, as well claiming a state commuter tax credit. Additional motivations for employers include reduced employee parking supply, sustainability and recognition, and improved physical wellness of employees (*13*).

Many success stories of such transit incentive programs have been reported. Reviews of some of them are as follows:

Pucher (*14*) explored the distributional effect of transit subsidy benefits and tax burdens among multiple income classes at the national level and found that transit subsidization does redistribute income from affluent households to low-income households. Pucher (*15*) also confirmd that the poor are significant beneficiaries of transit subsidies not only at the national level, but also in six U.S. metropolitan areas.

Meyer and Beimborn (*16*) presented a case study of a discounted pass program at the University of Wisconsin-Milwaukee and evaluated the program's impact on mode choice, transit ridership, and other travel-related issues, such as emissions and vehicle miles traveled. Overall, based on data from mail surveys and telephone interviews, they concluded that the transit pass program is highly effective in promoting transit use, and such a program benefits both users and non-users.

Hirsch et al. (*3*) studied the effects of various fare incentives, such as free intermodal transfers, express bus fare reduction, and discounts on bulk purchase, on New York City transit ridership. They concluded that fare incentives create greater-than-expected transit ridership increases.

Ecola and Grant (*17*) used interviews and other survey data from seven U.S. public transit agencies to examine the effects of transit benefit programs on ridership, revenues, and costs. They found that these benefit programs can help increase ridership and revenues, while decreasing costs.

Zhou and Schweitzer (*1*) presented a case study of a transit incentive program called "Dump the Pump" in Los Angeles, California. Eligible employees could exchange their parking permits for a free transit pass in a 12-week trial period, and a discounted transit pass after the expiration date. They found this program was attractive to drivers in the long run and more drivers would switch to transit if fuel prices are relative high and the travel time difference between transit and auto is relatively small.

Bueno et al. (*18*) investigated to what extent commuter benefits can incentivize changes in daily transportation habits, especially mode choice, in the States of New York and New Jersey. Data from the 2010–2011 Regional Household Travel survey were involved in the study and a multinomial logit model was used. Their results showed that employer-funded benefit programs for transit passes can significantly boost public transit use.

Although the empirical studies reviewed above have demonstrated the positive impacts of various transit incentive programs, none of them are concerned with designing such programs. For example, should incentives be flat or differentiated? If flat, what is the optimal amount? A shortcoming within the majority of those empirical studies is that financial incentives are assumed to be costless. Only the benefits of those benefits programs are analyzed and their costs are underestimated, because the effect of costly public funds is neglected.

Theoretical Studies

As reviewed in Sun and Schonfeld (*10*), public transit subsidization has been controversial for decades, although it is a worldwide practice to finance public transit systems in cases of operating deficits and to attract transit users by applying financial incentives. The most cited rationale for this practice is the so-called "Mohring effect" (see Gómez-Lobo (*4*) for a recent review), which states that passengers' costs are lower if more passengers use the transit system, because demand growth creates more vehicle runs and also makes transit routes and stops denser. Elgar and Kennedy (*19*) provided other rationales, such as positive externalities and second-best pricing.

Some recent studies about public transit subsidization include the following:

Larsen (*20*) studied the design of the Norwegian incentive contracts for public transportation. Under such a contract, decisions on the level of service—such as service frequencies—are left to the operators, considering that a public transportation operator has the best knowledge of the local market's demand. Fares are regulated by a public transit authority. A proper remuneration scheme (where operators receive fixed remuneration rates per revenue km and vehicle hour for basic and additional peak services, respectively) was used to ensure the operators are providing an optimal level of service from a social point of view.

Zou and Mizokami (*8*) proposed a performance-based incentive subsidy scheme where the bus operator is motivated to improve the service level and also reduce deficits. The motivation for the operators in this scheme is a premium from the local government, awarded for reduced operating deficits.

Sun and Schonfeld (*9*) presented a bilevel optimization model to study the behavior interactions between a profit-driven operator and the government. Various subsidization and regulation schemes were compared; their results showed that the combination of regulation and subsidization yields the highest social welfare and that subsidization schemes, without regulation, can produce socially worse results than the benchmark model.

Sun et al. (*10*) improved the model proposed by Sun and Schonfeld (*9*) by considering the effect of costly public funds. Their analytical and numerical results showed that the cost of public funds determines how the passengers' surplus is sacrificed to recover the fixed part of the operating cost. When the cost of public funds exceeds the critical shadow price of the financial constraints, subsidies are not warranted.

Although the models reviewed above provide important insights into the public transit subsidization and incentivization from a theoretical point of view, some critical issues have not yet been addressed. For example, most studies assume the operators obtain subsidies and very few are concerned with designing user incentives. More importantly, demand modeling is greatly sacrificed to obtain closed-form solutions, such as in the studies produced by Sun and Schonfeld (*9*) and Sun et al. (*10*). Such analyses are at the system level, and riders are usually homogeneous.

This study extends the existing theoretical economic literature on public transit incentivization by modeling a more realistic demand pattern and customizing incentives to different rider groups.

Microeconomic Model

A transit line covering a commuter corridor, as illustrated in Figure 1, has a set of stops, $N = \{0, 1, 2, \cdots, m\}$, and each stop is denoted as i or $j \in N$. Two links, namely a_i^e and a_i^w, are associated with a transit stop i except for the initial one, Stop 0. A set of links is denoted as $A = \{a_i^e, a_i^w | i \in N \backslash \{0\}\}$ and a link is denoted as $a \in A$. An origin–destination (OD) pair is denoted as $d \in D = \{(i,j)| i,j \in N, i \neq j\}$. If a link $a \in A$ is used by an OD pair $d \in D$, parameter $\phi_{ad} = 1$; otherwise, $\phi_{ad} = 0$. It is clear that if a link and an OD pair have different directions of travel, $\phi_{ad} = 0$. The set of traveler groups is G and each group is denoted as $g \in G$.

While we can consider a continuous distribution of value of time, it becomes challenging to customize incentives to each passenger group because the number of passenger groups may be infinite (assuming that each group of passengers has a unique value of time). Therefore, it is more practical to consider a limited number of passenger groups, such as two or three. In such a case, passengers could be potentially grouped by their incomes or other socio-demographic characteristics.

For each OD pair $d \in D$ and each rider group $g \in G$, a generalized cost (or travel impedance) p_{dg} is defined, which is a weighted sum of transit vehicle headway h, access and egress time X_d, in-vehicle travel time T_d, and fare f.

$$p_{dg} = f + e_{wg}\gamma h + e_{xg}X_d + e_{tg}T_d, \forall d \in D, g \in G \quad (1)$$

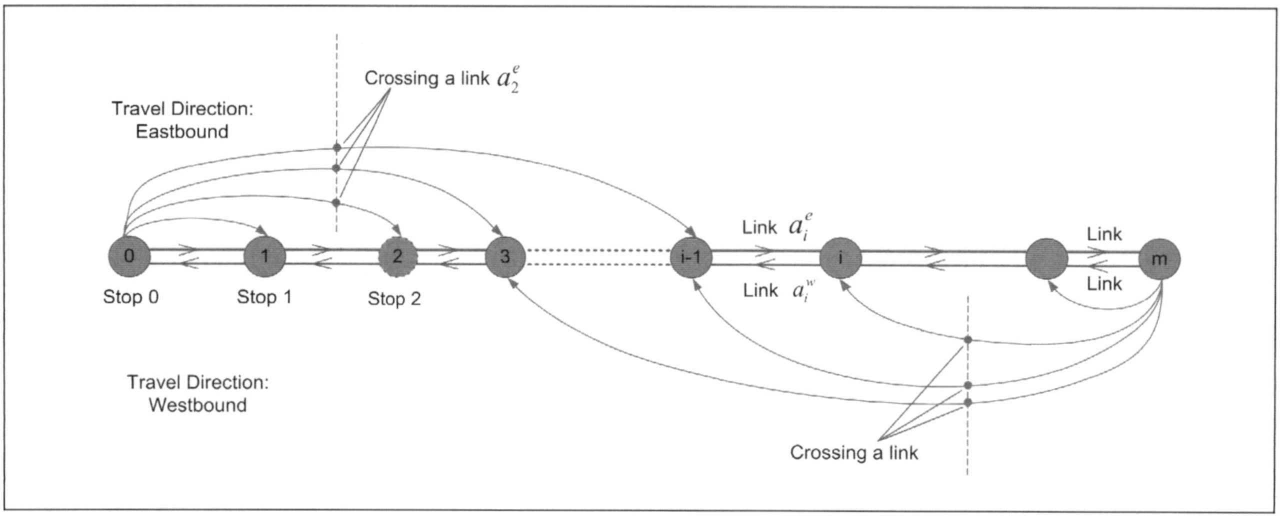

Figure 1. Illustration of a transit route and a many-to-many demand pattern.

where e_{wg}, e_{xg}, e_{tg} are values of time (VOT) for waiting time, access and time, and in-vehicle travel time, for group g, respectively. The typical 0.5 value of γ in the literature is obtained by assuming random passenger arrivals, sufficient vehicle capacities, and a constant headway.

Without loss of generality, it is assumed that the government can customize the incentive amount to each group. A monetary incentive w_g is transferred to a rider in group g after a transit trip is completed by that rider. Note that this formulation does not preclude the possibility that all groups receive the same incentive, i.e., $w_1 = w_2 = \cdots = w_g, \forall g \in G$. For each OD pair d and each group g, the adjusted travel cost after incentives is therefore:

$$\tilde{p}_{dg} = p_{dg} - w_g, \ \forall d \in D, \ g \in G \tag{2}$$

To consider demand elasticity, a linear demand function of the adjusted generalized cost is considered for each OD pair and rider group:

$$q_{dg} = Q_{dg}(1 - e_{pg}\tilde{p}_{dg}), \ \forall d \in D, \ g \in G \tag{3}$$

where q_{dg} is the actual demand, Q_{dg} is the potential demand, and e_{pg} is a coefficient measuring how demand is sensitive to travel cost. When $\tilde{p}_{dg} = \frac{1}{e_{pg}}$, $q_{dg} = 0$, meaning there are no transit trips.

OD passenger flows can be aggregated over links to obtain the link demand.

$$L_a = \sum_{d \in D} \phi_{ad} \sum_{g \in G} q_{dg}, \ \forall a \in A \tag{4}$$

The total supplier cost is hourly cost per vehicle multiplied by fleet size. Hourly cost per vehicle b is considered as a function of vehicle capacity k, namely:

$$b = \alpha + \beta k \tag{5}$$

where α is fixed hourly cost and β is marginal hourly cost. This linear cost structure is to model the economics of vehicle size. For example, the fixed part represents the driver's salary, which can be regardless of the number of passenger seats.

Since fleet size is round-trip time T divided by headway h, the total supplier cost is written as:

$$c = (\alpha + \beta k)\frac{T}{h} \tag{6}$$

The operator's payoff (profit) is as follows:

$$\pi = f \sum_{d \in D} \sum_{g \in G} q_{dg} - c \tag{7}$$

The total user surplus for an OD pair $d \in D$ and a group $g \in G$ is calculated by integrating the demand function over the adjusted generalized cost δ from $\tilde{p}_{dg}$ to $\tilde{p}_{0,g}$. $\tilde{p}_{0,g}$ is the adjusted travel impedance that drives actual demand q_{dg} to be zero, i.e., $\tilde{p}_{0,g} = \frac{1}{e_{pg}}$, $g \in G$.

$$U_{dg} = \int_{\tilde{p}_{dg}}^{\tilde{p}_{0,g}} q_{dg}(\delta) d\delta, \ \forall d \in D, \ g \in G \tag{8}$$

The government, with an objective of social welfare maximization, determines the amount of incentive per transit trip to be provided to riders and designs the fare and headway limits, i.e., the highest fare level f_m and the maximum allowable headway h_m. The operator, which can observe all decisions made by the government, maximizes its own profit by selecting fare, headway, and vehicle capacity.

The government thus solves the following optimization problem:

$$\max_{\{f_m, h_m, w_g\}} \sum_{d\in D}\sum_{g\in G}((f-(1+\theta)w_g)q_{dg}) - (\alpha+\beta k)\frac{T}{h} + \sum_{d\in D}\sum_{g\in G}U_{dg} \tag{9}$$

subject to

$$f\sum_{d\in D}\sum_{g\in G}q_{dg} - (\alpha+\beta k)\frac{T}{h} \geq \varepsilon \tag{10}$$

$$f_m \geq 0 \tag{11}$$

$$h_m \geq 0 \tag{12}$$

$$w_g \geq 0,\ \forall g \in G \tag{13}$$

In objective (9), θ is the average cost of raising per-unit public funds. As discussed in Sun et al. (*10*), the government's funds are mainly tax revenues. Due to the distortionary impact of taxation, one tax dollar costs more than one. If on average, 1.1 units of social welfare are lost due to the distortion of resource allocation, when only one unit of public fund is raised, the parameter $\theta = (1.1-1.0)/1 = 0.1$. Sun et al. (*10*) provided extended discussions on the origin and magnitude of the cost of public funds. Constraint (10) is the individual rationality constraint to ensure that the operator is willing to operate and the minimum profit is ε, which is a nonnegative constant.

The operator optimizes fare, headway, and vehicle capacity, subject to the capacity constraint and policy constraints. Constraint (15) ensures that the accumulated demand in one headway can be served by one vehicle. Constraints (16) and (17) specify that neither fare nor headway can exceed the policy limit determined by the government.

The operator solves the following problem:

$$\max_{\{f,h,k\}} f\sum_{d\in D}\sum_{g\in G}q_{dg} - (\alpha+\beta k)\frac{T}{h} \tag{14}$$

subject to

$$h\sum_{d\in D}\phi_{ad}\sum_{g\in G}q_{dg} \leq k,\ \forall a \in A \tag{15}$$

$$0 \leq f \leq f_m \tag{16}$$

$$0 \leq h \leq h_m \tag{17}$$

$$k \geq 0 \tag{18}$$

As analyzed in Sun and Schonfeld (*9*), under the symmetric information assumption, at optimality the government would specify f_m and h_m in a way that $f = f_m$ and $h = h_m$. By contradiction, if $f < f_m$ or $h < h_m$, the government can always reduce f_m or h_m until $f = f_m$ and $h = h_m$, while not sacrificing its objective (9).

Since policy constraints are binding, i.e., $f = f_m$ and $h = h_m$, f_m and h_m can be removed from the formulation. After rearrangements, the government's problem and the operator's problem can be combined and converted into a nonlinear program, which can be rewritten as follows:

$$\min_{\{f,h,k,w_g\}} \sum_{d\in D}\sum_{g\in G}\left(((1+\theta)w_g - f)Q_{dg}\left(1-e_{pg}\tilde{p}_{dg}\right)\right) - \sum_{d\in D}\sum_{g\in G}\left(\int_{\tilde{p}_{dg}}^{\tilde{p}_{0,g}} Q_{dg}(1-e_{pg}\delta)d\delta\right) + (\alpha+\beta k)\frac{T}{h} \tag{19}$$

subject to

$$(\alpha+\beta k)\frac{T}{h} + \varepsilon - f\sum_{d\in D}\sum_{g\in G}Q_{dg}(1-e_{pg}\tilde{p}_{dg}) \leq 0 \tag{20}$$

$$h\sum_{d\in D}\phi_{ad}\sum_{g\in G}Q_{dg}(1-e_{pg}\tilde{p}_{dg}) - k \leq 0,\ \forall a \in A \tag{21}$$

$$-f \leq 0 \tag{22}$$

$$-h \leq 0 \tag{23}$$

$$-k \leq 0 \tag{24}$$

$$-w_g \leq 0,\ \forall g \in G \tag{25}$$

Note that $\tilde{p}_{dg}$ is a function of f, w_g, and h. The integral $\int_{\tilde{p}_{dg}}^{\tilde{p}_{0,g}} Q_{dg}(1-e_{pg}\delta)d\delta$ can be simplified into $Q_{dg}\left(\frac{1}{2e_{pg}} - \tilde{p}_{dg} + \frac{e_{pg}}{2}\left(\tilde{p}_{dg}\right)^2\right)$.

Analytical Solution

The Karush-Kuhn-Tucker (KKT) conditions for this problem are:

Stationary conditions:

$$\sum_{d\in D}\sum_{g\in G}((f-(1+\theta)w_g)Q_{dg}e_{pg}) + \mu_1 f\sum_{d\in D}\sum_{g\in G}(Q_{dg}e_{pg}) - \mu_1\sum_{d\in D}\sum_{g\in G}\left(Q_{dg}\left(1-e_{pg}\tilde{p}_{dg}\right)\right) - \sum_{a\in A}\mu_2^a h\sum_{d\in D}\phi_{ad}\sum_{g\in G}Q_{dg}e_{pg} - \mu_f = 0 \tag{26}$$

$$\sum_{d\in D}\sum_{g\in G}\left(((1+\mu_1)f-(1+\theta)w_g)Q_{dg}e_{pg}\frac{e_{wg}}{2}\right)$$
$$+\sum_{d\in D}\sum_{g\in G}\left(Q_{dg}\frac{e_{wg}}{2}\left(1-e_{pg}\tilde{p}_{dg}\right)\right)$$
$$-(1+\mu_1)(\alpha+\beta k)\frac{T}{h^2} \quad (27)$$
$$+\sum_{a\in A}\mu_2^a\sum_{d\in D}\phi_{ad}\sum_{g\in G}Q_{dg}\left(1-e_{pg}\tilde{p}_{dg}\right)$$
$$-\sum_{a\in A}\mu_2^a h\sum_{d\in D}\phi_{ad}\sum_{g\in G}Q_{dg}e_{pg}\frac{e_{wg}}{2}-\mu_h=0$$

$$(1+\mu_1)\frac{\beta T}{h}-\sum_{a\in A}\mu_2^a-\mu_k=0 \quad (28)$$

$$\sum_{d\in D}\left(\theta Q_{dg}\left(1-e_{pg}\tilde{p}_{dg}\right)+((1+\theta)w_g-(1+\mu_1)f)Q_{dg}e_{pg}\right)$$
$$+\sum_{a\in A}\mu_2^a h\sum_{d\in D}\phi_{ad}\sum_{g\in G}Q_{dg}e_{pg}-\mu_w^g=0,\ \forall g\in G \quad (29)$$

Complementary slackness:

$$\mu_1\left((\alpha+\beta k)\frac{T}{h}+\varepsilon-f\sum_{d\in D}\sum_{g\in G}Q_{dg}(1-e_{pg}\tilde{p}_{dg})\right)=0 \quad (30)$$

$$\mu_2^a\left(h\sum_{d\in D}\phi_{ad}\sum_{g\in G}Q_{dg}(1-e_{pg}\tilde{p}_{dg})-k\right)=0,\ \forall a\in A \quad (31)$$

$$-\mu_f f=0 \quad (32)$$

$$-\mu_h h=0 \quad (33)$$

$$-\mu_k k=0 \quad (34)$$

$$-\mu_w^g w_g=0,\ \forall g\in G \quad (35)$$

where μ_1, μ_2^a, μ_f, μ_h, μ_k, and μ_w^g are nonnegative multipliers for constraints (20–25), respectively. In practice, headway h and vehicle capacity k are strictly positive, namely $h>0$ and $k>0$. Then, $\mu_h=0$ and $\mu_k=0$ from Equations 33 and 34.

From Equation 28, we obtain:

$$(1+\mu_1)\frac{\beta T}{h}=\sum_{a\in A}\mu_2^a \quad (36)$$

Denote the link with the highest demand, i.e., $\max_{a\in A}\{h\sum_{d\in D}\phi_{ad}\sum_{g\in G}Q_{dg}(1-e_{pg}\tilde{p}_{dg})\}$, as $\bar{a}$. Then, we analyze two mutually exclusive cases.

Case 1 (Capacity constraint for link $\bar{a}$ is not binding.):

If the capacity constraint is unbinding for link $\bar{a}$, namely:

$$\sum_{d\in D}\phi_{\bar{a}d}\sum_{g\in G}Q_{dg}(1-e_{pg}\tilde{p}_{dg})<\frac{k}{h} \quad (37)$$

by Equation 31, $\mu_2^{\bar{a}}=0$. Obviously, for all other links $a\in A\backslash\{\bar{a}\}$, their capacity constraints are not binding, because $\sum_{d\in D}\phi_{ad}\sum_{g\in G}Q_{dg}\ (1-e_{pg}\tilde{p}_{dg})<\max_{a\in A}\{h\sum_{d\in D}\phi_{ad}\sum_{g\in G}Q_{dg}(1-e_{pg}\tilde{p}_{dg})\}$ and $\mu_2^a=0,\ \forall a\in A\backslash\{\bar{a}\}$. Then, $\sum_{a\in A}\mu_2^a=0$, thus contradicting Equation 36. Note that multiplier $\mu_1\geq 0$ and $(1+\mu_1)\frac{\beta T}{h}>0$.

Case 1 is then excluded.

Case 2 (Capacity constraint for link $\bar{a}$ is binding.):

The capacity constraint is binding for link $\bar{a}$:

$$\sum_{d\in D}\phi_{\bar{a}d}\sum_{g\in G}Q_{dg}(1-e_{pg}\tilde{p}_{dg})=\frac{k}{h} \quad (38)$$

For other links with demand lower than $\sum_{d\in D}\phi_{\bar{a}d}\sum_{g\in G}Q_{dg}(1-e_{pg}\tilde{p}_{dg})$, their capacity constraints cannot be binding, i.e., $\sum_{d\in D}\phi_{ad}\sum_{g\in G}Q_{dg}(1-e_{pg}\tilde{p}_{dg})<\frac{k}{h}$, $\forall a\in A\backslash\{\bar{a}\}$ and $\mu_2^a=0,\ \forall a\in A\backslash\{\bar{a}\}$. Then, to make $\sum_{a\in A}\mu_2^a$ positive, $\mu_2^{\bar{a}}$must be positive.

To summarize, Equation 36 means only the highest link demand is used to determine the vehicle capacity and the associated multiplier $\mu_2^{\bar{a}}>0$.

Inserting Equation 36 into Equation 26 yields:

$$\sum_{d\in D}\sum_{g\in G}\left(\left((1+\mu_1)f-(1+\theta)w_g-\mu_1\left(\frac{1}{e_{pg}}-\tilde{p}_{dg}\right)-(1+\mu_1)\beta T\phi_{\bar{a}d}\right)Q_{dg}e_{pg}\right)=\mu_f \quad (39)$$

To satisfy $\mu_f\geq 0$, f can only be positive. If $f=0$, it is clear that the left-hand side of Equation 39 is negative, contradicting the nonnegativity of μ_f. Since $f>0$, due to the complementary slackness (32), $\mu_f=0$. We can rewrite Equation 39 into:

$$\sum_{d\in D}\sum_{g\in G}\left(\left(f-\frac{1+\theta}{1+\mu_1}w_g-\frac{\mu_1}{1+\mu_1}\left(\frac{1}{e_{pg}}-\tilde{p}_{dg}\right)-\beta T\phi_{\bar{a}d}\right)Q_{dg}e_{pg}\right)=0 \quad (40)$$

For each OD pair d and each group g, λ_{dg} is used to denote $f - \frac{1+\theta}{1+\mu_1} w_g - \frac{\mu_1}{1+\mu_1}\left(\frac{1}{e_{pg}} - \tilde{p}_{dg}\right) - \beta T \phi_{\bar{a}d}$. If there is only one OD pair and one traveler group, namely $|D| = 1$ and $|G| = 1$, we have a single value of λ, as follows:

$$\lambda = f - \frac{1+\theta}{1+\mu_1} w - \frac{\mu_1}{1+\mu_1}\left(\frac{1}{e_p} - \tilde{p}\right) - \beta T \quad (41)$$

To satisfy Equation 40, λ must be zero. Therefore, Equation 41 can be used as a pricing formula when demand is homogeneous. When there are many OD pairs and multiple traveler groups, it is not necessarily that $\lambda_{dg} = 0, \forall d \in D, g \in G$. As analyzed in the next section, for some combinations of OD pair and traveler group, $\lambda_{dg} > 0$, while for some others, $\lambda_{dg} < 0$. This fact is referred to as "cross-subsidization." Those combinations with negative λ values are cross-subsidized by others with positive λ values and the weighted sum of all λ values as shown in Equation 40 is zero.

Another equation can be derived from the optimality condition about headway. Inserting Equation 36 into Equation 27 and noting Equation 38 yields:

$$\sum_{d \in D}\sum_{g \in G}\left(\left(f - \frac{1+\theta}{1+\mu_1} w_g + \frac{1}{1+\mu_1}\left(\frac{1}{e_{pg}} - \tilde{p}_{dg}\right) - \beta T \phi_{\bar{a}d}\right) Q_{dg} e_{pg} \frac{e_{wg}}{2} h\right) = \alpha \frac{T}{h} \quad (42)$$

Equation 42 establishes the relation between the waiting cost per OD pair and per group, i.e., $\frac{e_{wg}}{2} h$, and the fixed operating cost, i.e., $\alpha \frac{T}{h}$. If there is only one OD pair and one traveler group, Equation 42 is reduced into:

$$Q_{dg}\left(1 - e_{pg}\tilde{p}_{dg}\right)\frac{e_{wg}}{2} h = \alpha \frac{T}{h} \quad (43)$$

Equation 43 means the total passenger waiting cost should be equal to the fixed operating cost.

For each group, an equation can be obtained by inserting Equations 36 and 38 into Equation 29.

$$\sum_{d \in D}\left(\left(-f + \frac{1+\theta}{1+\mu_1} w_g + \frac{\theta}{1+\mu_1}\left(\frac{1}{e_{pg}} - \tilde{p}_{dg}\right) + \beta T \phi_{\bar{a}d}\right) Q_{dg} e_{pg}\right) = \frac{\mu_w^g}{1+\mu_1}, \ \forall g \in G \quad (44)$$

A summation of Equation 44 over $g \in G$ is:

$$\sum_{d \in D}\sum_{g \in G}\left(\left(-f + \frac{1+\theta}{1+\mu_1} w_g + \frac{\theta}{1+\mu_1}\left(\frac{1}{e_{pg}} - \tilde{p}_{dg}\right) + \beta T \phi_{\bar{a}d}\right) Q_{dg} e_{pg}\right) = \sum_{g \in G}\frac{\mu_w^g}{1+\mu_1} \quad (45)$$

Combining Equation 40 and 45 yields:

$$\sum_{d \in D}\sum_{g \in G}\left(\frac{\theta - \mu_1}{1+\mu_1}\left(1 - e_{pg}\tilde{p}_{dg}\right) Q_{dg}\right) = \sum_{g \in G}\frac{\mu_w^g}{1+\mu_1} \quad (46)$$

Since $\mu_w^g \geq 0, \forall g \in G$, it is clear that $\theta \geq \mu_1$. Numerical analyses in the next section show the relation between the cost of public funds θ and the shadow price for the individual rationality constraint μ_1 and how this relation impacts incentive decisions.

Numerical Studies

Model Inputs

While analytical equations are derived in the previous section, numerical analyses are conducted in this section to illustrate important relationships, with a single rail transit line consisting of eight stations as an example. The hourly OD matrix is shown in Table 1. It can be observed that the westbound travel (from a station with a larger index to a station with a smaller index, as illustrated in Figure 1) is heavier than the eastbound.

Three rider groups are considered and their fractions are 0.3, 0.3, and 0.4. For each OD pair, the in-vehicle travel time is specified in Table 2. Other parameters, such as VOTs and vehicle operating costs, are provided in Table 3. For e_{wg}, e_{ag}, e_{p_g}, and e_s, their three values are for three traveler groups, respectively. The minimum profit level $\varepsilon = 1000$ \$/hour. The value of ϕ_{ad} for each link a and each OD pair d can be found easily, as illustrated in Figure 1. For example, travelers from stops 0 and 1 heading to stops 2 and beyond must traverse link a_2^e. Due to space limits, values of ϕ_{ad} are not presented in this paper.

Results

The mathematical model is programmed and solved with GAMS (General Algebraic Modeling System). KNITRO (*21*) is used to solve the nonlinear program. The starting values of f, h, and k are 2.5, 0.1, and 200, respectively. Figure 2 plots how incentives and fares change with respect to the cost of public funds θ. As θ grows, incentives per trip decrease until θ reaches Threshold 1, where

Table 1. Hourly OD Demand (passengers/hour)

OD	0	1	2	3	4	5	6	7
0	0.0	107.5	177.8	238.4	273.3	283.7	246.1	193.3
1	144.1	0.0	160.9	266.9	343.1	377.5	336.7	270.8
2	305.8	206.5	0.0	221.8	365.5	439.0	405.9	335.7
3	456.0	380.8	246.6	0.0	289.3	412.2	402.6	346.0
4	591.3	553.9	459.8	327.3	0.0	263.4	304.4	286.8
5	686.2	681.3	617.4	521.3	294.5	0.0	164.7	201.5
6	704.4	718.9	675.4	602.4	402.6	194.9	0.0	121.7
7	665.8	695.8	672.3	623.1	456.6	286.9	146.5	0.0

Table 2. In-Vehicle Travel Time (minutes)

Time	0	1	2	3	4	5	6	7
0	0.0	2.8	5.1	7.2	10.0	12.9	15.0	17.7
1	2.8	0.0	2.3	4.4	7.2	10.1	12.2	14.9
2	5.1	2.3	0.0	2.1	4.9	7.8	9.9	12.6
3	7.2	4.4	2.1	0.0	2.8	5.7	7.8	10.5
4	10.0	7.2	4.9	2.8	0.0	2.9	5.0	7.7
5	12.9	10.1	7.8	5.7	2.9	0.0	2.2	4.8
6	15.0	12.2	9.9	7.8	5.0	2.2	0.0	2.6
7	17.7	14.9	12.6	10.5	7.7	4.8	2.6	0.0

Table 3. Baseline Values of Some Parameters

Symbols	Definitions	Baseline values	Unit
A	Average access & egress time	0.03	Hour
e_{wg}	Value of waiting time by group	20 / 25 / 30	\$/hour
e_{ag}	Value of access time by group	20 / 25 / 30	\$/hour
e_{p_g}	Coefficient by group	0.1 / 0.09 / 0.08	–
e_s	Value of in-vehicle time	8 / 10 / 12	\$/hour
T	Round-trip time	0.6	Hour
α	Fixed hourly cost of vehicle	1,500	\$/hr
β	Variable hourly cost of vehicle	8	\$/hr/seat

the incentive offered to Group 3 becomes zero. Incentives for the other two groups keep declining until Threshold 2 is reached. When θ is beyond Threshold 2, no incentives are provided and fare remains stable, which means it is too expensive to offer public funds to transit riders as incentives.

Figure 3 numerically illustrates Equation 46. First, the multiplier for constraint (30) is always below the 45º line, meaning $\mu_1 \leq \theta$. Second, when incentives for all three groups are positive, namely $w_g > 0, \forall g \in G$, by complementary slackness (35), $\sum_{g \in G} \mu_w^g = 0$. Then, by Equation

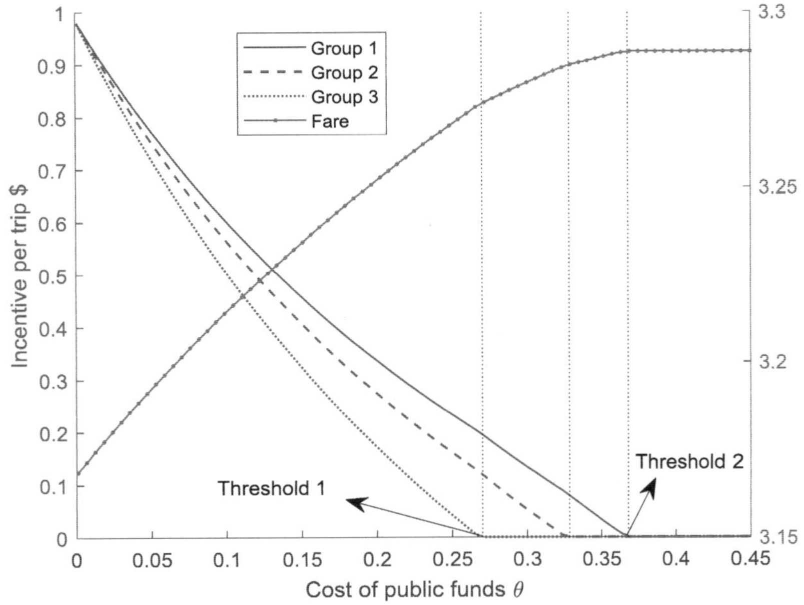

Figure 2. Optimized incentives and fare versus cost of public funds.

46, $\mu_1 = \theta$, corresponding to the fact that the curve for μ_1 overlaps with the 45º line within 0 and Threshold 1.

Beyond Threshold 1, Figure 3 shows riders in Group 3 are no longer eligible for incentives through public funds, which can also be explained by Equation 46. Figure 3 shows $\mu_1 < \theta$ as θ is beyond Threshold 1. When $\mu_1 < \theta$, by Equation 46, $\sum_{g \in G} \mu_w^g > 0$, which means at least one μ_w^g is positive, implying at least one group of riders lose incentives, by complementary slackness (35).

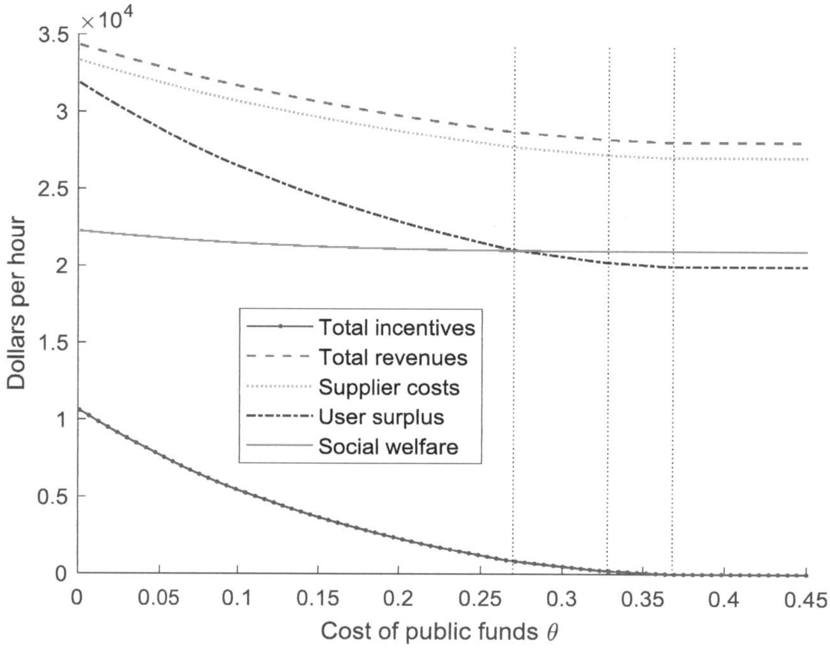

Figure 4. Changes of various costs and benefits with respect to cost of public funds.

Figure 4 shows how social welfare as well as other costs and benefits change with cost of public funds. First, as θ grows, the social welfare decreases, due to the increasing disruptive effect of taxation. Second, the difference between total revenues and supplier costs is fixed at the minimum profit level of \$1000/hour in the base example. This means that any extra profit beyond the minimum level is harmful to the society as a whole.

Sensitivity Analysis of Minimum Profit

A sensitivity analysis of the minimum profit ε is conducted and relevant results are shown in Figure 5. As ε increases, headway h and vehicle capacity k remain the same, and supplier costs are thus fixed according to Equation 6. As shown in Figure 5, fare f and incentives w_g increase linearly with ε at the same rate, which implies $f - w_g$ is fixed. Hence, for each combination of d and g, the right-hand side of Equation 47 is fixed.

$$\tilde{p}_{dg} = f - w_g + e_{wg}\gamma h + e_{xg}X_d + e_{tg}T_d,\ \forall d \in D,\ g \in G \tag{47}$$

Since the adjusted travel cost is fixed, according to Equation 3, the actual demand and user surpluses remain the same. Due to the fare increase, total revenues increase to keep the profit growing. In summary, to satisfy the minimum profit requirement, increasing incentives are applied, but all of the increments are transferred to the operator in order to increase profit by charging a higher fare.

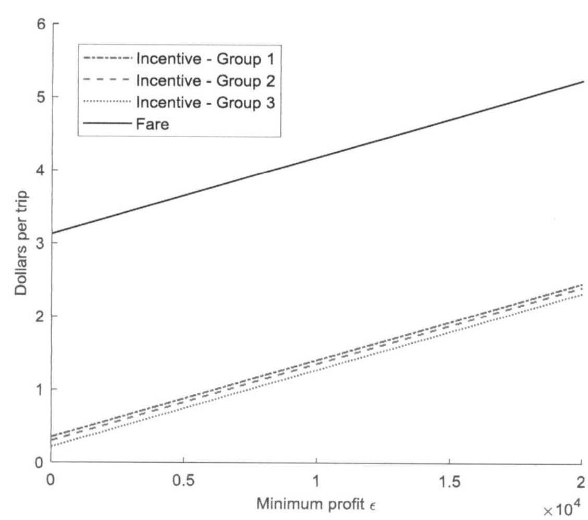

Figure 5. Changes of incentives and fare with minimum profit.

Since the money transfer is costly, as ε grows, the social welfare decreases. When the growth in ε is one

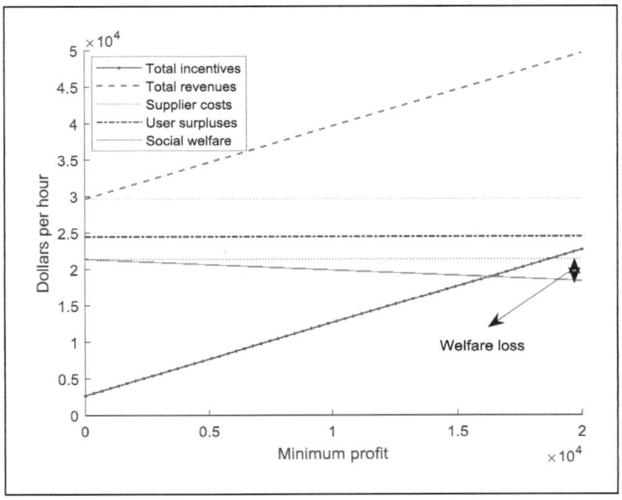

Figure 6. Sensitivity analysis of the minimum profit.

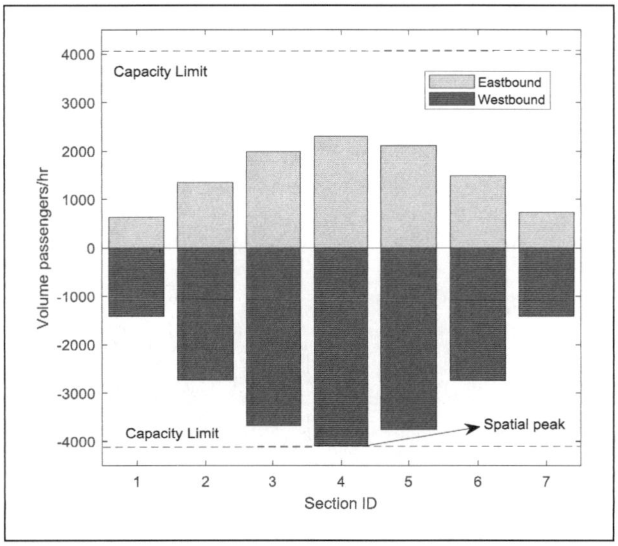

Figure 7. Aggregated actual demand for each link.

unit, the social welfare loss is θ units, as illustrated in Figure 6.

Cross-Subsidization

In a transportation system where travelers have several feasible origins and destinations, the utilization of transportation facilities is seldom uniform over space, which gives rise to peaks of aggregate travel needs, even in a linear transportation corridor. Figure 7 shows the aggregated actual demand for each link in the eastbound direction (defined in Figure 1). A negative value means demand in the opposite (i.e., westbound) direction. It is obvious that, although different links face quite different passenger volumes, once the highest demand need is satisfied, the supplied capacity is sufficient at all other links. One important implication is that only travelers who directly contribute to peak link volume should be responsible for the transit operating cost, because the capacity is designed to satisfy the peak demand and the operating cost depends on the capacity. For other travelers, such as those who travel eastbound, even if they travel a longer distance, their increases in travel will not directly affect the capacity design, as long as the increases do not form a new demand peak (with an even higher volume) in the eastbound direction.

Table 4 presents the value of λ_d for each OD pair $d \in G$. By definition, $\lambda_d = \sum_{g \in G} \lambda_{dg}, \forall d \in G$. Then, $\lambda_d = \sum_{g \in G} \left(f - \frac{1+\theta}{1+\mu_1} w_g - \frac{\mu_1}{1+\mu_1} \left(\frac{1}{e_{pg}} - \tilde{p}_{dg} \right) - \beta T \phi_{\bar{a}d} \right)$, which describes how passengers from different OD pairs are cross-subsidizing each other. An OD pair d with a negative λ_d is cross-subsidized by other OD pairs with positive λ_d values. For example, passengers of those OD pairs in the left-bottom corner of Table 4 use the bottleneck link, namely "Link 4-Westbound", and they jointly contribute to the highest link volume, which determines the amount of supplied capacity. Since those OD pairs use "Link 4-Westbound", $\phi_{\bar{a}d} = 1$ for those OD pairs. From the formula of λ_d, it is not difficult to find that $\lambda_d < 0$.

Conclusions

A microeconomic model is proposed for the optimal design of public transit incentive programs. After reviewing relevant empirical studies and theoretical models on public transit subsidy or incentive programs, several gaps were identified in the literature. First, a majority of studies assume the transit operators are the only recipients of subsidies. Second, analyses are usually conducted at the system level and the demand pattern is greatly simplified; for instance, all travelers have the same origin and destination. Third, homogeneous VOT and cost elasticities are assumed. Therefore, in this study, a many-to-many travel pattern, multiple rider groups, and user-based incentives are considered in the economic model.

Important analytical relations are derived from the optimality conditions of the optimization model. Numerical analyses are also used to illustrate some important findings from this study:

(1) Depending on the magnitude of the cost of public funds, all riders are eligible for the monetary incentives, or only certain rider groups are eligible, or none of them are eligible.
(2) The shadow price of the individual rationality constraint is bound by the cost of public funds

Table 4. Matrix of Lambda Values

λ value	0	1	2	3	4	5	6	7
0	0.0	19.7	33.3	45.7	53.8	57.4	50.9	41.0
1	26.4	0.0	29.3	49.7	65.7	74.4	67.8	55.9
2	57.3	37.6	0.0	40.3	68.4	84.5	79.9	67.8
3	87.3	70.8	44.8	0.0	52.9	77.7	77.6	68.5
4	−131.7	−126.4	−107.0	−77.5	0.0	48.2	57.0	55.2
5	−149.0	−151.7	−140.2	−120.5	53.9	0.0	29.9	37.7
6	150.0	157.0	150.5	136.7	75.4	35.4	0.0	22.2
7	138.3	148.4	146.4	138.2	87.9	53.6	26.8	0.0

and when all riders receive incentives, the shadow price equals the cost of public funds.

(3) Sensitivity analyses show that as the minimum profit level increases, more financial incentives are applied while the rider surplus remains the same, which jointly contributes to the decline of social welfare.

(4) When a many-to-many travel pattern creates demand peaks, the effect of cross-subsidization is identified, with those travelers who directly contribute to the capacity bottleneck being subsidized by others.

The efficiency loss from distortionary taxation depends on the type of taxes and varies from one jurisdiction to another (*22*). In this study, it is assumed that public funds are costly due to the deadweight loss from taxes. In certain cases, taxation can create positive impacts. For example, progressive taxation has a positive effect on reducing income inequality because a larger share of the income is collected from the rich than from the less wealthy. Such positive impacts should be weighed against the efficiency losses (*23*).

The proposed model can be used by public transit agencies, major employers, or universities to design transit incentive programs, with consideration of costly public funds and traveler heterogeneity. The present study can be improved in the following aspects:

(1) Although significant contributions are made in this paper, especially regarding the modeling of a many-to-many demand pattern and demand heterogeneity, several other aspects of public transit operations are simplified. For example, only a single line is studied and only one time period is considered. Future studies can improve upon the current model by considering these important but neglected characteristics of public transit operations.

(2) Although a many-to-many demand pattern is considered, which enables the capture of the spatial demand peak, future studies should also introduce demand dynamics so that the temporal demand peak is also incorporated.

(3) While baseline values of some parameters are used in this study, other model users are encouraged to calibrate the parameters (e.g., operating cost and value of time) of the proposed model with their own data.

Acknowledgments

The financial support from the U.S. Department of Energy through its ARPA-E's TRANSNET Program is acknowledged.

Author Contributions

The authors confirm contribution to the paper as follows: study conception and design: Yanshuo Sun (YS), Lei Zhang (LZ); data collection: YS; analysis and interpretation of results: YS, LZ; draft manuscript preparation: YS. Both authors reviewed the results and approved the final version of the manuscript.

References

1. Zhou, J., and L. Schweitzer. Getting Drivers to Switch: Transit Price and Service Quality Among Commuters. *Journal of Urban Planning and Development*, Vol. 137, No. 4, 2011, pp. 477–483.
2. Myers, G., D. Hagen, T. Russo, C. McMullin, A. Lembrick, B. Silbaugh, and K. Parker. Benefits of Campus Transit Pass: Study of Students' Willingness to Pay for Proposed Mandatory Transit Pass Program. *Transportation Research Record: Journal of the Transportation Research Board*, 2006. 1971: 133–139.
3. Hirsch, L., J. Jordan, R. Hickey, and V. Cravo. Effects of Fare Incentives on New York City Transit Ridership.

Transportation Research Record: Journal of the Transportation Research Board, 2000. 1735: 147–157.

4. Gómez-Lobo, A. Monopoly, Subsidies and the Mohring Effect: A Synthesis. *Transport Reviews*, Vol. 34, No. 3, 2014, pp. 297–315.
5. Van Reeven, P. Subsidisation of Urban Public Transport and the Mohring Effect. *Journal of Transport Economics and Policy*, Vol. 42, No. 2, 2008, pp. 349–359.
6. Savage, I., and K. A. Small. A Comment on "Subsidisation of Urban Public Transport and the Mohring Effect." *Journal of Transport Economics and Policy*, Vol. 44, No. 3, 2010, pp. 373–380.
7. Basso, L. J., and S. R. Jara-Díaz. The Case for Subsidisation of Urban Public Transport and the Mohring Effect. *Journal of Transport Economics and Policy*, Vol. 44, No. 3, 2010, pp. 365–372.
8. Zou, W., and S. Mizokami. Incentive Subsidy Scheme Design with Elastic Transport Demand. *Journal of Advanced Transportation*, Vol. 48, No. 8, 2014, pp. 927–941.
9. Sun, Y., and P. M. Schonfeld. Optimization Models for Public Transit Operations Under Subsidization and Regulation. *Transportation Research Record: Journal of the Transportation Research Board*, 2015. 2530: 44–54.
10. Sun, Y., Q. Guo, P. Schonfeld, and Z. Li Implications of the Cost of Public Funds in Public Transit Subsidization and Regulation. *Transportation Research Part A: Policy and Practice*, Vol. 91, 2016, pp. 236–250.
11. TranServe, U.S. Department of Transportation. https://www.transportation.gov/transerve/. Washington, D.C. Accessed July 15, 2017.
12. Commuter Choice Maryland, Maryland Transit Administration. http://www.commuterchoicemaryland.com/. Hanover, Maryland. Accessed July 15, 2017.
13. Saelens, B. E., A. V. Moudon, B. Kang, P. M. Hurvitz, and C. Zhou. Relation Between Higher Physical Activity and Public Transit Use. *American Journal of Public Health*, Vol. 104, No. 5, 2014, pp. 854–859
14. Pucher, J. Equity in Transit Finance: Distribution of Transit Subsidy Benefits and Costs Among Income Classes. *Journal of the American Planning Association*, Vol. 47, No. 4, 1981, pp. 387–407.
15. Pucher, J. Who Benefits from Transit Subsidies? Recent Evidence from Six Metropolitan Areas. *Transportation Research Part A: General*, Vol. 17, No. 1, 1983, pp. 39–50.
16. Meyer, J., and E. Beimborn. Usage, Impacts, and Benefits of Innovative Transit Pass Program. *Transportation Research Record: Journal of the Transportation Research Board*, 1998. 1618: 131–138.
17. Ecola, L., and M. Grant. Impacts of Transit Benefits Programs on Transit Agency Ridership, Revenues, and Costs. *Journal of Public Transportation*, Vol. 11, No. 2, 2008, pp. 1–18.
18. Bueno, P. C., J. Gomez, J. R. Peters, and J. M. Vassallo. Understanding the Effects of Transit Benefits on Employees' Travel Behavior: Evidence from the New York-New Jersey Region. *Transportation Research Part A: Policy and Practice*, Vol. 99, 2017, pp. 1–13.
19. Elgar, I., and C. Kennedy. Review of Optimal Transit Subsidies: Comparison Between Models. *Journal of Urban Planning and Development*, Vol. 131, No. 2, 2005, pp. 71–78.
20. Larsen, O. I. "Designing Incentive Schemes for Public Transport Operators in Hordaland County, Norway." Presented at the 7th International Conference on Competition and Ownership in Land Passenger Transport (Thredbo), Molde, Norway, 2001.
21. Knitro, A. *Nonlinear Optimization Solver*. https://www.artelys.com/en/optimization-tools/knitro. Accessed July 15, 2017.
22. Barrios, S., J. Pycroft, and B. Saveyn. *The Marginal Cost of Public Funds in the EU: The Case of Labour versus Green Taxes (No. 35)*. Directorate General Taxation and Customs Union, European Commission. Luxembourg: Publications Office of the European Union, 2013, pp. 1–48.
23. Bjart, H., and G. H. Bjertnæs. *The Size of the Marginal Cost of Public Funds: A Discussion with Special Relevance to Norway*. Reports 2015/13. Statistisk sentralbyrå (Statistics Norway), Oslo, Norway, 2015.

The Standing Committee on Transportation Economics (ABE20) peer-reviewed this paper (18-00735).

Sage Publications USA
sagepub.com

ISBN 978-1-5108-7731-3

9 781510 877313

CURRAN ASSOCIATES INC.
proceedings.com